ERROR AND UNCERTAINTY
IN SCIENTIFIC PRACTICE

History and Philosophy of Technoscience

Series Editor: *Alfred Nordmann*

Forthcoming

Experiments in Practice
Astrid Schwarz

ERROR AND UNCERTAINTY
IN SCIENTIFIC PRACTICE

EDITED BY

Marcel Boumans, Giora Hon and Arthur C. Petersen

Routledge
Taylor & Francis Group

LONDON AND NEW YORK

First published 2014 by Pickering & Chatto (Publishers) Limited

Published 2016 by Routledge
2 Park Square, Milton Park, Abingdon, Oxfordshire OX14 4RN
711 Third Avenue, New York, NY 10017, USA

First issued in paperback 2015

Routledge is an imprint of the Taylor & Francis Group, an informa business

BRITISH LIBRARY CATALOGUING IN PUBLICATION DATA

Error and uncertainty in scientific practice. – (History and philosophy of
technoscience)
1. Research – Methodology. 2. Research – Management. 3. Errors, Scientific –
Prevention.
I. Series II. Boumans, Marcel editor of compilation. III. Hon, Giora editor of
compilation. IV. Petersen, Arthur, editor of compilation.
507.2-dc23

ISBN-13: 978-1-138-66227-8 (pbk)
ISBN-13: 978-1-8489-3416-0 (hbk)

Typeset by Pickering & Chatto (Publishers) Limited

CONTENTS

ACKNOWLEDGEMENTS

Most of the contributions to this volume were originally discussed at the workshop, 'Error in the Sciences', which took place from 24 to 28 October 2011 at the Lorentz Center in Leiden, the Netherlands. The workshop was co-sponsored by PBL Netherlands Environmental Assessment Agency and the *Netherlands Institute for Advanced Study in the Humanities and Social Sciences* (NIAS). We would like to thank the Lorentz Center and in particular Henriette Jensenius, Sietske Kroon and Mieke Schutte, for their financial and excellent organizational support. We are very grateful to all the participants of this workshop for their contributions to the discussions of the various papers, without whom this workshop would not have been so fruitful: Jonathan Barzilai, Bruce Beck, Frans Birrer, Evelyn Forget, Victor Gijsbers, Richard Gill, Gabriele Gramelsberger, Maarten Hajer, Maria Jimenez Buedo, Bart Karstens, Joel Katzav, Luca Mari, Deborah Mayo, James McAllister, Ave Mets, Mary Morgan, Wendy Parker, Julian Reiss, Jutta Schickore, Leonard Smith, Aris Spanos, Kent Staley, Jacob Stegenga, Eran Tal, David Teira Serrano, Lex van Gunsteren and Ernst Wit.

LIST OF CONTRIBUTORS

M. Bruce Beck is Wheatley-Georgia Research Alliance Professor and Eminent Scholar in Environmental Systems and Water Quality in the Warnell School of Forestry and Natural Resources, University of Georgia. Since 1990 he has worked closely with the US Environmental Protection Agency on issues of model validation, model quality assurance and the analysis of uncertainty. He is editor of the 2002 book *Environmental Foresight and Models: A Manifesto* (New York: Elsevier).

Yakov Ben-Haim is Professor of Mechanical Engineering and holds the Yitzhak Moda'i Chair in Technology and Economics at the Technion–Israel Institute of Technology. He initiated and developed info-gap decision theory for modelling and managing severe uncertainty. Info-gap theory is applied in engineering, biological conservation, economics, project management, climate change management, homeland security, medicine and other areas (see info-gap.com). He has been a visiting scholar in many countries around the world and has lectured at universities, medical and technological research institutions, and central banks. He has published more than ninety articles, and five books.

Marcel Boumans is Associate Professor of History and Philosophy of Economics at the University of Amsterdam and the Erasmus University Rotterdam. His research is marked by three Ms: modelling, measurement and mathematics. His main research focus is on understanding empirical research practices in economics from (combined) historical and philosophical perspectives. On these topics he has published a monograph, *How Economists Model the World into Numbers* (London: Routledge, 2005); edited the volume *Measurement in Economics: A Handbook* (New York: Elsevier, 2007); co-authored a textbook, *Economic Methodology: Understanding Economics as a Science* (Basingstoke and New York: Palgrave Macmillan, 2010); and co-edited the volume *Histories on Econometrics* (Durham, NC, and London: Duke University Press, 2011).

Alessandro Giordani is a Senior Lecturer at the Department of Philosophy, Catholic University of Milan, where he teaches courses in logic and philosophy of science. In addition to his work on the foundation of measurement and the role of models in measurement and science, his research topics include modal logic, epistemic logic, epistemic foundation and the application of logical tools for analysing the process of knowledge.

Giora Hon, Professor of History and Philosophy of Science at the University of Haifa, Israel, published widely on the concept of error in science and philosophy. His edited book with Jutta Schickore and Friedrich Steinle, *Going Amiss in Experimental Research*, appeared in 2009 (Springer). For his recent work (with Bernard R. Goldstein) on modelling, see 'Maxwell's Contrived Analogy: An Early Version of the Methodology of Modeling', *Studies in History and Philosophy of Modern Physics*, 43 (2012), pp. 236–57.

Bart Karstens is Lecturer in Philosophy of Science at Leiden University. He is a member of the NOW-sponsored research programme, 'Philosophical Foundations of the Historiography of Science'. His research interests include the history of the humanities and the philosophy of historiography. His publications on approaches to the study of past science include 'Bopp the Builder. Discipline Formation as Hybridization: The Case of Comparative Linguistics', in R. Bod, J. Maat and T. Weststeijn (eds), *The Making of the Humanities. Volume II: From Early Modern to Modern Disciplines* (Amsterdam: Amsterdam University Press, 2012), pp. 103–27 and 'Towards a Classification of Approaches to the History of Science', *Organon*, 43 (2011), pp. 5–28.

Luca Mari, PhD in Measurement Science, is Professor at the Cattaneo University – LIUC, Castellanza, Italy. He is currently Chairman of TC1 (Terminology) and Secretary of TC25 (Quantities and Units) of the International Electrotechnical Commission (IEC), IEC expert in WG2 (VIM) of the Joint Committee for Guides in Metrology (JCGM), and former Chairman of TC7 (Measurement Science) of the International Measurement Confederation (IMEKO). He is the author or co-author of several scientific papers on measurement science, published in international journals and international conference proceedings.

Deborah G. Mayo is Professor in the Department of Philosophy at Virginia Tech and holds a visiting appointment at the Centre for the Philosophy of Natural and Social Science at the London School of Economics. She is the author of *Error and the Growth of Experimental Knowledge* (Chicago, IL: University of Chicago Press, 1996), which won the 1998 Lakatos Prize, awarded to the most outstanding contribution to the philosophy of science during the previous six years. Professor Mayo co-edited the volume *Acceptable Evidence: Science and Values in Risk Management* (New York and Oxford: Oxford University Press, 1991, with R. D. Hollander) and *Error and Inference* (Cambridge: Cambridge University Press, 2010, with A. Spanos). She has published numerous articles on the philosophy and history of science and on the foundations of statistics and experimental inference and interdisciplinary works on evidence relevant for regulation and policy. She is currently writing a book on how to tell what is true about statistical inference.

Arthur C. Petersen is Chief Scientist at the PBL Netherlands Environmental Assessment Agency, Professor of Science and Environmental Public Policy at the VU University Amsterdam, Visiting Professor at the London School of Economics and University College London, and Research Affiliate at MIT. He studied physics and philosophy, obtained PhD degrees in atmospheric sciences and philosophy of science, and now also finds disciplinary homes in sociology and political science. Most of his research is about managing uncertainty.

Leonard A. Smith received his PhD (Physics) from Columbia University. He is a Professor of Statistics at the London School of Economics and directs the Centre for the Analysis of Time Series. Since 1992 he has also been a Senior Research Fellow (Mathematics) of Pembroke College, Oxford. His research focuses on nonlinear dynamical systems, predictability, the role of probability in decision support, and the implications uncertainty, ambiguity and ignorance hold when relating mathematical results to reality. In 2003 he received the Royal Meteorological Society's Fitzroy Prize for his contributions to applied meteorology.

Aris Spanos is Wilson Schmidt Professor of Economics at Virginia Tech. He obtained his PhD at the London School of Economics in 1982. He is the author of two textbooks in econometrics and more than seventy papers in refereed journals in economics, econometrics, statistics, economic methodology and philosophy of science. More recently, he co-edited a book with Deborah G. Mayo on *Error and Inference* (Cambridge: Cambridge University Press, 2010). Specific areas of current research include philosophical and technical problems pertaining to statistical modelling and inference, model validation, securing the reliability and precision of inference and the empirical modelling of financial and macro data.

Kent W. Staley is Associate Professor of Philosophy at Saint Louis University. His work focuses on issues surrounding experimental practice and reasoning, drawing largely on the history of experimental physics as well as statistical theory and inferential practice. He is the author of *The Evidence for the Top Quark: Objectivity and Bias in Collaborative Experimentation* (Cambridge: Cambridge University Press, 2004).

LIST OF FIGURES AND TABLES

INTRODUCTION

Marcel Boumans and Giora Hon

There is no general theory of error, and there will never be one. Indeed, the task of theorizing error is insurmountable. The author of the entry, 'Erreur', in the *Encyclopédie* of Diderot and d'Alembert had already cautioned in the mid-eighteenth century that, although several philosophers detailed the errors of the senses, the imagination and the passions, their imperfect theories are ill-suited to cast light on practical decisions. And, he continued, the imagination and passions are enfolded in so many ways and depend so strongly on temperaments, times and circumstances, that it is impossible to uncover all the hidden forces that the imagination and passions activate.[1] A century later the English logician de Morgan concurred, 'There *is* no such thing as a classification of the ways in which men may arrive at an error; it is much to be doubted whether there ever *can be*'.[2] The study of error cannot commence with a successful stroke of disambiguation upon which a coherent and all-embracing theory could be founded.[3] Historically and etymologically 'error' may be traced to the Latin root 'errare', which originally had two contrasting meanings: first, 'to go this way and that, to walk at random'; and second, 'to go off the track, to go astray'. So right at the origin of the term, there is a tension between aimless wandering and straying from some definite path.[4] Either way the metaphor is spatial, expressing the unknown features of *terra incognita* versus the informative map of *terra firma*.[5]

The figure of 'The Error' in Coypel's allegorical painting, *Truth Unveiled by Time*, produced around 1702, offers a vivid illustration of the epistemic predicament of error. The painting depicts a blindfolded man standing alone, curious, anxious to see and to touch, his hands restless, groping under a veil of darkness. In his *Memoirs of the Blind*, Derrida commented on this allegory of Error, the searching blindfolded man:

> *Naturally* his eyes *would be able* to see. But they are *blindfolded* (by a handkerchief, scarf, cloth, or veil, a textile, in any case, that one fits over the eyes and ties behind the head). Erect and blindfolded, not naturally but by the hand of the other, or by his own hand, obeying a law that is not natural or physical since the knot behind the head remains within a hand's reach of the subject who could undo it: it is as if the subject of the error had consented to having got it up, over his eyes, as if he got off on his suf-

fering and his wandering, as if he chose it, at the risk of a fall, as if he were playing at seeking the other during a sublime and deadly game of blind man's buff.[6]

The allegory questions the origin of the predicament of error: whose hand blindfolded the wanderer and for what reason? According to Derrida, it is not Nature's hand. Derrida thus positioned himself in the long tradition of philosophy and science, where one of the fundamental ingrained presuppositions is that Nature neither errs nor deceives. As famously formulated by Einstein, 'Subtle is the Lord, but malicious He is not'.

In contrast to the Aristotelian position, we commonly hold with Newton that 'Nature ... is ... always consonant to itself'.[7] Put differently in Galileo's words, 'the errors ... lie ... not in geometry or physics, but in a calculator who does not know how to make a true accounting'.[8] In other words, nature never errs; still, as Francis Bacon observed, it is 'like a labyrinth, where on all sides the path is so often uncertain, the resemblance of a thing or a sign is deceptive, and the twists and turns ... are so oblique and intricate'.[9] But Bacon nourished hopes; he was confident that his *New Organon* would help '[to lay] the foundations not of a sect or of a dogma, but of human progress and empowerment'. For he rejected 'all that hasty human reasoning, based on preconceptions, which abstracts from things carelessly and more quickly than it should, as a vague, unstable procedure, badly devised'. Bacon expected in light of his method, that men would

> put off the zeal and prejudices of beliefs and think of the common good; then, freed
> from obstacles and mistaken notions of the way, and equipped with our helps and
> assistance, we would ask them to undertake their share of the labours that remain.[10]

In 1787 Kant picked up this thread and opened the second edition of his *Critique of Pure Reason* with this claim of Bacon as a motto: 'the end of unending error'.[11]

Nature is consonant to itself; errors are therefore the product of human thought and action. There is thus another reason why there is no general theory of error, and there will never be one. This has to do with what theories can do in offering explanations of phenomena. According to Bogen and Woodward,[12] well-developed scientific theories predict and explain facts about phenomena but not facts about the data which are the raw material of evidence for these theories. Many different factors play a role in the production of any bit of data, and the characteristics of such factors are heavily dependent on the peculiarities of, for example, the particular experimental design, detection devices, or data-gathering procedures which the investigator applies. Data are idiosyncratic to particular experimental contexts, and typically cannot occur outside these contexts. Indeed, the elements involved in the production of data will often be so disparate and numerous, and the details of their interactions so complex, that it is impossible to construct a theory that would allow us to predict their occurrence or trace in detail how they combine to produce specific bits of data. Phenomena, by contrast, are not idiosyncratic to specific experimental contexts. Bogen and Woodward argue that phenomena have stable, repeatable characteristics which can be detectable by means of a variety of different procedures; however, these procedures may yield in different contexts quite different data. Bogen and Woodward go further and claim that, in fact, no theory is needed for dealing with the data:

> an important source of progress in science is the development of procedures for the systematic handling of observational and measurement error and procedures for data-analysis and data-reduction which obviate the need for a theory to account of what is literally seen.[13]

Errors are no phenomena; rather, they are idiosyncratic, case-specific and local misapprehensions. Addressing them will differ from case to case and depend upon the effects of many different conditions peculiar to the subject under investigation, such as the experimental design and the equipment used. 'The factors contributing to observational error, and the means for correcting for it in different cases, are far too various to be captured in an illuminating way in a single general account.'[14]

Undoubtedly, 'the senses, the imagination, and the passions' block the path to knowledge, but here we witness the very execution of one of the central pieces of the method of science, namely experimentation, having the potential to lead astray, and there is no general theory that will guide us.

This is not to say that there have never been attempts at taming error. The Theory of Error, that is, the mathematical theory of error, has a long and rich history based on the pioneering works of illustrious mathematicians such as Legendre, Laplace and Gauss.[15] With key elements such as the least squares, the central limit theorem and the normal distribution, the 'wanderings' are brought back to the beaten tracks. But the theory could be built only on abstractions. It arose when observational errors could be divorced from their causes, from idiosyncratic circumstances, from the individual observer, indeed from actual measurements. The history of the development of the theory of error is a history of creating objectivity: the elimination of personal judgement by 'mechanical' rules of calculation.[16]

The root of this approach may be traced to the controversy between Galileo and Chiaramonti in the *Dialogue Concerning the Two Chief World Systems*.[17] The controversy offers an early view of the fundamentals of error theory. Galileo states that

> the principal activity of pure astronomers is to give reasons just for the appearances of celestial bodies, and to fit to these and to the motions of the stars such a structure and arrangement of circles that the resulting calculated motions correspond with those same appearances.

Thus, the general, motivating assumption is that 'if observations ... were correct, and if the calculations ... were not erroneous, both the former and the latter would necessarily have to yield exactly the same distance'. The underlying assumption is that the two procedures, namely the observation and the calculation, should correspond exactly. Nevertheless, as Galileo reports of Chiaramonti's calculations of the distance of the nova of 1572 from earth, not even two calculations came out in agreement. This is the fate of the astronomer, 'however well the astronomer might be satisfied merely as a calculator, there was no satisfaction and peace for the astronomer as a scientist'. Simplicio, one of the protagonists of the dialogue, is forced then to accept that

> all [of Chiaramonti's calculations] were fallacious, either through some fault of the computer or some defect on the part of the observers. At best I might say that a single one, and no more, might be correct; but I should not know which one to choose.

It is in this context that Galileo – while offering his own calculations – develops a rudimentary theory of error, some kind of a preliminary statistical analysis of astronomical observations that facilitates the right choice of observations and calculations that lead finally to a reliable result.

Galileo's theory of error appears to be directly linked to the objective certainty he apprehends in mathematical propositions which is due to the necessity of their truth and its correspondence to physical reality. Thus, Salviati instructs Simplicio that 'astronomers and mathematicians have discovered infallible rules of geometry and arithmetic, by means of which ... one may determine the distance of the most sublime bodies within one foot'. Under the guidance of Salviati, Simplicio surmises that the differences in the estimates of the altitude of the new star of 1572 were due not to a defect in the rules of geometry and arithmetic but to 'errors made in determining ... angles and distances by instrumental observations'.

Galileo clearly realized that it is hard to achieve accuracy, 'both on account of the imperfection of astronomical instruments, which are subject to much variation, and because of the shortcomings of those who handle them with less care than is required'. Elsewhere in the *Dialogue*, Galileo underlines 'the possible occurrence of errors in the instrumental observations themselves because the observer is not able to place the center of the pupil of his eye at the pivot of the sextant'. Galileo goes further in recognizing that 'the very instruments of seeing introduces a hindrance of its own'.

Galileo's belief that mathematics, geometry and arithmetic is the source of certainty in the natural sciences and especially in astronomy underlies his insight concerning the occurrence, distribution and estimation of error in observation. On the basis of this mathematical analysis of observational error, Galileo reaches the conclusion that 'everything supports the opinion of those who place [the nova] among the fixed stars'. In Galileo's view his analytical procedure which undermines Chiaramonti's calculations that render the nova a sublunar phenomenon, demonstrates 'more clearly and with ... much greater probability' that the nova should be placed in the most remote heavens.

Hald summarized the principles underlying Galileo's analysis of observational errors as follows:

- [True value] There is just *one number* which gives the distance of the star from the centre of the earth, the *true distance*.
- [Sources of error] *All* observations are encumbered with errors, due to the observer, the instruments and other observational conditions.
- [Distribution] The observations are distributed *symmetrically* about the true value; that is, the *errors are distributed symmetrically about zero*.
- [The nature of the distribution] *Small errors occur more frequently than large errors.*
- [Propagation] As the calculated distance is a *function of the direct angular observations*, errors influence the calculated distances so that small adjustments of the observations may give large adjustments of the distances.[18]

Admittedly, Galileo applied these principles rather intuitively and did not formalize them. Nevertheless, his refutation of Chiaramonti's conclusion is a *tour de force* of integrating mathematical argumentation with physical assessment of possible sources of errors that may arise in the use of instrumentation for astronomical observations – the motivating principle being that 'what happens in the concrete ... happens the same way in the abstract'. The mathematical scientist, *filosofo geometra*, has to conduct a true accounting if he wants 'to recognize in the concrete the effects which he has proved in the abstract'. On all accounts Galileo was thinking in terms of objective certainty which we can now contrast with modern concepts such as 'subjective probability', or 'degree of belief'; the latter concepts of course play no part in this procedure. Astronomers had not adopted this method of data assessment until Boscovich made use of it more than a century later.[19]

However, even a century later there was no realization that errors should form an important part of all observational and experimental procedures. Lambert's attempt to find a general theory of errors was thus exceedingly novel. In 1765 he took a wide range of physical problems as examples, and he grouped them together according to similarity in the type of error treatment they required rather than according to the similarity of the phenomena they treated. Lambert was one of the pioneers who considered the problem of observational and experimental error in its own right.

The nineteenth century explodes with new developments, especially the consolidation of error theory and the spread of probability theory to new domains. To turn on the fast-forward mode we hear now of Simpson and Bayes, the extension of the calculus of probability to the social sciences, Quetelet, Bessel and the personal equation and thence Fechner and the development of psychophysics as a prelude to experimental psychology, till we reach the development of heredity with Galton, who experimentally demonstrated the normal law.[20]

The mix of probability as a mathematical theory with experiment led to much confusion with respect to the validity of the normal law, a confusion to which Lippmann wittily referred in his remark to Poincaré: 'Everybody believes in the exponential law of errors: the experimenters, because they think it can be proved by mathematics; and the mathematicians, because they believe it has been established by observation'.[21]

The distribution of errors follows the normal law only approximately, even when the quantity to be measured is as steady as possible. Although this approximation can be justified under much wider conditions, namely the central limit theorem, it is still the case, as Jeffreys remarks, that these conditions 'are seldom *known* to be true in actual applications'. 'Consequently', Jeffreys adds, 'the normal law as applied to actual observations can be justified, in the last resort, only by comparison with the observations themselves'.[22] Needless to say, this is a vicious circle; it is the result of justifying the treatment of observations by exclusively referring to the observations themselves.

In the National Physical Laboratory's *Code of Practice* it is stated that 'there is of course no reason for experimental observations to follow the normal distribution exactly – it is a convenient mathematical expression which fits most of the experimental observations'. The *Code* stresses that 'it should be recognized that this is an assumption which may not always be justified'.[23] Indeed, as Margenau critically remarks,

> experience presents the scientist with innumerable *skew* distributions, differing perceptibly from the Normal law. These he often dismisses or corrects, because for some hitherto unstated reason he objects to them. He uses the normal distribution both as an inductive generalization from experience and as a criterion for the trustworthiness of that experience. Thus he is lifting himself by his bootstraps unless an independent argument can be given for the normalcy of that distribution.[24]

The correct approach is 'to regard the number following the plus-or-minus sign as an estimate of the width parameter of some', and it should be stressed, *some* 'statistical distribution of observed values which would be obtained if the measurement were replicated a number of times'.[25] Clearly, the appeal to probability is an attempt to break the vicious circle.

Moreover, there are fields and disciplines where a statistical mean of observations is meaningless or lacks relevance. In economics, discussing errors in social statistics, Morgenstern was highly critical about the mindless usage of the theory of error:

The world would, indeed, be even more of a miracle than it is if the influence of one set of errors offsets that of another set of errors so conveniently that we need not to bother much with the whole matter.[26]

Although the theory of error became the standard theory to deal with errors across many different disciplines, its legitimacy depends on the availability of a large number of independent observations, equally trustworthy so far as skill and care are concerned, and obtained with precise instruments. These requirements cohere with the science of astronomy, but there are many scientific disciplines where these requirements can never be met.

The theory of observational and experimental error is based on the classification which distinguishes between two categories of error, namely, systematic and random. Systematic errors are defined as errors that *could be* measured and calculated for each observation, given the values of certain parameters. In other words, systematic errors do not lend themselves to probabilistic treatment. Evidently, random errors do not have this feature. Typically, it is claimed that the skilled experimenter can eliminate all systematic errors, and upon completing this task the experimenter finds, so the claim goes, that there is still a margin of error which requires further consideration. This margin of error, it is asserted, is due to random errors. Thus, the experimenter is culpable when he or she commits, so to speak, systematic errors. The experimenter may be blamed for failing to eliminate systematic errors or to account for them, but he or she may not be blamed for the occurrence of random errors; the latter are inevitable and the experimenter is not liable for their occurrence. It is perhaps due to this clear-cut distinction between culpability and inevitability that this classification has become so attractive to experimenters.[27]

Note that the dichotomy between systematic and random errors does not focus on the source of the error; rather, it examines the nature of the error by applying a mathematical criterion. This criterion judges whether the estimation of the error is derived by a statistical analysis of repeated measurements or by non-statistical methods in which much depends on the judgement of the experimenter as well as measurements and calculations in allocating limits to the accuracy of the measurement. The former error is random, the latter systematic.

The concept of uncertainty as a quantifiable attribute is relatively new in the history of measurement. It was in the early 1980s that official codes of practices introduced 'uncertainty measurement' to cover multiple usage of the term 'error' which the theory of error does not address. The concepts of error and error analysis have long been a part of measurements in science, engineering and metrology. When all of the known or suspected components of an error have been evaluated, and the appropriate corrections have been applied, an uncertainty still remains about the 'truthfulness' of the stated result, that is, a doubt about how well the result of the measurement represents the 'value' of the quantity being measured. The expression 'true value' is not used in this procedure since the true value of a measurement may never be known. But a 'true value' is invoked in error analysis. Thus, the result of an analysis after correction may by chance be very close to the value of the measurand, and hence have a negligible error. However, the uncertainty may still be very large, simply because the analyst is very unsure of how close that result is to the value.

In a *Code of Practice* formulated by NPL, the National Physical Laboratory, the uncertainty of a measurement is divided into two categories: the random uncertainty and the systematic uncertainty. According to this *Code*, the estimation of random uncertainty is derived by a statistical analysis of repeated measurements while the estimation of systematic uncertainty is assessed by non-statistical methods and much depends on the judgement of the experimenter in allocating limits to this uncertainty. Here we reach the source of the distinction between error and uncertainty measurement.

In 1885 Edgeworth published his paper, 'Observations and Statistics', in the *Philosophical Magazine*. In this essay on the theory of errors of observation and the first principle of statistics he drew the following distinction:

> Observations and statistics agree in being quantities grouped about a Mean; they differ, in that the Mean of observations is real, of statistics is fictitious. The mean of observations is a cause, as it were the source from which diverging errors emanate. The mean of statistics is a description, a representative quantity put for a whole group, the best representative of the group, that quantity which, if we must in practice put one quantity for many, minimizes the error unavoidably attending such practice. Thus measurements by the reduction of which we ascertain a real time, number, distance are observations. Returns of prices, exports and imports, legitimate and illegitimate marriages or births and so forth, the averages of which constitute the premises of practical reasoning, are statistics. In short observations are different copies of one original; statistics are different original affording one 'generic portrait'. Different measurements of the same man are observations; but measurements of different men, groups round l'homme moyen, are primâ facie at least statistics.[28]

According to Stigler, the historian of statistics,

> Edgeworth's aim was to apply the tools developed in the previous century for *observations* in astronomy and geodesy, where a more or less objectively defined goal made it possible to quantify and perhaps remove nonmeasurement error and meaningful to talk of the remaining variation as random error, to social and economic *statistics* where the goal was defined in terms of the measurements themselves and the size and character of the variation depended upon the classification and subdivisions employed.[29]

This distinction between error theory and statistics in general is critical. The theory of probability and statistics evolved parallel to the development of the theory of observational and experimental error with the result of sometimes blurring the dividing line. And this brings us to the fundamental assumptions of the *Guide to the Expression of Uncertainty in Measurement*, namely:

(1) 'It is possible to characterize the *quality* of a measurement by accounting for both systematic and random errors *on a comparable footing*'.[30] A method is provided, using the concept of measurement uncertainty, for casting the information previously obtained from 'error analysis' into probabilistic terms. Quality is thereby turned into a measurement of quantity.

(2) 'It is not possible to state how well the essentially unique true value of the measurand is known, but only how well it is believed to be known'. Measurement uncertainty is 'a measure of how well one believes one knows the essentially unique true value of the measurand. The notion of "belief" is an important one, since it moves metrology into a realm where results of measurement need to be considered and quantified in terms of *probabilities* that express degrees of belief'.[31]

To be sure, beliefs are quantitative but, as Good put it, 'this does not mean that they are necessarily numerical; indeed, it would be absurd to say that your degree of belief that it will rain tomorrow is 0.491336230 ... Degrees of belief are only partially ordered'.[32] Clearly, the plus-or-minus interval of the theory of error is a totally well-ordered set. There arose, then, the need for a new measurement, namely, the measurement of uncertainty which is partially ordered.

Modern measurement science distinguishes between two types of evaluation of uncertainty, where uncertainty means that there is some doubt about the validity of a measurement result. Type A evaluation of uncertainty is an evaluation by statistical analysis of a series of observations that is based on the theory of error. Type B evaluation of uncertainty is an evaluation by means other than the statistical analysis of the observations. This type of evaluation investigates the conceptual/theoretical levels, the realizations, the representational power of observations/data/sampling, knowledge of environment of the phenomenon being studied, personal bias of the researcher and the reliability of the instruments/procedure/techniques, in other words aspects that cannot be analysed in a 'mechanical' way, as a routine or as a purely mathematical exercise. Type B evaluations are therefore less objective, and depend on the understanding, critical analysis, intellectual honesty and professional skill of those who contribute to these evaluations.

To sum up, the treatment of errors which arise in observations, measurements and experiments has a long history, beginning in the seventeenth century. This treatment makes use of various mathematical techniques with the goal of reaching quantitative (that is, numerical) values for correcting the result and, furthermore, assessing it. In recent years, another measure, uncertainty measurement, has been defined. It surprisingly employs similar techniques but reaches a different, categorically different numerical assessment which cannot be used for correction; it expresses degrees of belief, or subjective probability. A general account of error should take account, then, of these two aspects, namely, observational errors and evaluations of sources of errors that are not 'mechanical'. It is befitting to end on a remark which Margenau made half a century ago with regard to the theory of error: 'the philosopher of science is obliged to take note of this remarkable fact: both "truth" and "tolerance" must be fished out of the uncertainties of the immediately given by more or less arbitrary rules not immediately presented in Nature'.[33] Margenau made this remark about error analysis in 1950; we may say the same today with perhaps greater astonishment about uncertainty measurements.

The aim of this edited volume is to discuss various practices of dealing with errors and uncertainties in an attempt to attain reliability and thereby to gain a deeper understanding of what error and uncertainty in science and their treatment entail. While the daily practice of empirical research, in and outside the laboratory, is dominated by dealing with all kinds of errors and uncertainties seeking reliable results, there exists no general cross-disciplinary framework for dealing with these concepts. Various sophisticated procedures for the systematic handling of observational and measurement errors as well as procedures for data analysis and the associated uncertainties were and still are being developed, but they are all fragmented and mainly developed to address specific epistemological and methodological problems within a particular scientific domain. The reason that a more general account was lacking is that the kind of error to be corrected differs from case to case and depends upon the effects of many different conditions peculiar to the subject under investigation, the research design, and the equipment and/or models used: the issue is context dependent and field specific. The various practices of dealing with errors and uncertainties have developed their own separate methods and techniques, with little cross-fertilization. While these different methods are not likely to be integrated, more general solutions to their common problem, namely, how to take account of reliability, have been developed and are discussed in this volume. To be sure, contextual knowledge is not easily transmittable to different scientific domains, but methods for achieving reliability have an overarching feature.

The volume contains case studies of research practices across a wide variety of scientific and practical activities, and across a range of disciplines, including experimental physics, econometrics, environmental science, climate science, engineering, measurement science and statistics, with the aim of comparing different epistemologies and methodologies of treatments of error in various scientific practices. Such a comparison is achieved through cross-disciplinary transfer of diagnosis, prognosis and rectifying measures. We need to explore, on the one hand, practices of dealing with error and uncertainty in a specific scientific domain so we need practitioners to provide us with such details; we need, on the other hand, to generalize from these detailed case studies and make comparisons between different practices, and this is the work of philosophers and historians of science. We have therefore asked scientists working in various scientific domains and disciplines as well as philosophers and historians of science to discuss strategies of dealing with error and uncertainty in theory and in practice.

Like any goal-oriented procedure, experiment is subject to many kinds of error. They have a variety of features, depending on the particulars of their sources. The identification of error, its source, its context and its treatment shed light on practices and epistemic claims. Understanding an error amounts, inter alia, to uncovering the knowledge generating features of the system involved. The chapters will address the conceptualization of error in the historiography of science (Karstens) and experimental knowledge in the face of theoretical errors (Staley). Against this background, an extensive discussion of learning from error will be developed (Mayo).

Typically, we consider measurements reports of our knowledge of the state of the system under examination. Recently, measurement has shifted from a truth-seeking process to a model-based one in which the quality of the measurement is assessed by pragmatic aims. As a result of this epistemological shift, the quality of measurements are not reported in terms of accuracy, an expression which indicates the closeness of the result to the true value, but in terms of uncertainty. The shift has had implications on calibration strategies: instead of expecting that reference values are true, they are required only to be traceable. Three kinds of shifts will be explored (Mari and Giordani): (1) the shift from error to uncertainty; (2) the shift from accuracy assessment to quality assessment, and (3) the shift from standards as prototypes to standards as instrumental set-ups. Essentially, the theme will be the transition from error to uncertainty in measurement sciences.

Collaboration of scientists with decision-makers about complex models of great importance (e.g. energy, climate and finance) which contain large measures of uncertainty, has recently become acute. Interdisciplinary work has been done in this domain to arrive at a commonly agreed upon typology of uncertainties. This includes efforts to deepen understanding of reliability of complex modelling, since it is often not possible to establish the accuracy of the results of simulations or to quantitatively assess the impacts of different sources of uncertainty. In this section, recourse is made to qualitative assessment of the different elements comprising any research, that is, data, models, expert judgements and the like. Effort is made to determine 'methodological reliability', given the purpose of the relevant model in the interface between scientific and public domains. The chapters focus on the handling of uncertainty in models at the science–policy interface (Beck), and the role of the 'relevant dominant uncertainty' in science and in science-based policy, especially with regard to climate policy (Smith and Petersen). In general, it is surprising how we can and do manage in an environment not entirely known to us. We acknowledge our ignorance but do address the unknown. The info-gap theory facilitates analysis where science and policy meets on future risky projects (Ben-Haim).

Data of social science and statistics are typically inhomogeneous: as the realizations of complex interactions, they are not stable. Since the traditional statistical techniques presuppose homogeneity, they cannot be applied in these instances. Various 'ometrics'–disciplines arose as new branches of applied statistics by developing strategies for treating this kind of data. The new strategies share the feature of being model based. An evaluation of errors therefore is a model-based assessment, where the model must cover the sources of errors. Several strategies are discussed where errors are evaluated by the assessment of their representations as the process of learning from data (Spanos).

1 THE LACK OF A SATISFACTORY CONCEPTUALIZATION OF THE NOTION OF ERROR IN THE HISTORIOGRAPHY OF SCIENCE: TWO MAIN APPROACHES AND THEIR SHORTCOMINGS

Bart Karstens[1]

Introduction

One would expect that the concept of error figures large in the historiography of science. After all, history is studied because the past differed from the present. With respect to the history of science it is natural to suggest that such differences come about in differences between right and wrong. What else would be the incentive to change our theories about the world if these are never found to stand in need of correction? This insight is not new to historians but the strange thing is that it has not led to a satisfactory theory of error. As a matter of fact, talk of error in the traditional epistemic way has almost completely faded into the background in present-day historiography. This chapter aims to clarify this situation.

In the chapter two basic outlooks on the phenomenon of error will be discerned. The first I have named the 'errors as obstacles' approach. In this approach, errors are conceptualized as obstacles to progress and science is seen as a process in which errors are sifted out. This happens mainly through the application of rational procedures such as the use of scientific methods, principles of reasoning and so forth. From the second perspective, errors are basically seen as products of negotiation. I will refer to this group as the 'error as failures' approach. From this perspective, standards of truth and rationality are not operative in any transcendental sense but the efficacy of evaluative standards must be understood in their specific local, historical context. They are considered as the outcomes of social factors, such as negotiation. What counts as error is a derivative of these (temporary) outcomes. Claims to knowledge that have not succeeded and are seen as erroneous from the viewpoint of the ruling standards have failed to gain prominence. The conceptualization of error from the second perspective, then, is in terms of failures of gaining prominence instead of in terms of epistemic obstacles to progress.

The two main approaches to error thus have a completely different outlook on what errors are. The difference between them lies in the treatment of factors that are seen as determinants of truth and error in science. In the first approach, truth and error are explained in an asymmetrical manner. Truth must be accounted for by an appeal to rational methods. The occurrence of error is due to other, 'external' causes, which are mostly social factors. In the second approach no such asymmetries are in use any more. What is seen as truth and error, i.e. the acceptance and rejection of all knowledge claims, are the outcomes of similar processes of negotiation. Thus the same type of factors are considered to be constitutive of all knowledge claims, whether they are held true or held false does not matter for the mode of explanation.

It is important to note that the distinction between the two main approaches to error that is drawn in this chapter is not drawn across the line of realism versus anti-realism. In various guises both positions can be found in each of the two groups. The line *is* drawn in terms of determining factors in science. Any approach to past science in which at some point a distinction between internal and external factors is made will find itself listed under the main 'errors as obstacles' approach. When an internal realm of science is discerned this is mostly done by attributing a special character to scientific rationality. Any approach that is based on moving beyond the internal–external distinction finds itself under the main 'errors as failures' approach. In the latter case no determining factor is credited with some special character: rational and social factors are considered to be deeply intertwined. It follows that the same *type* of factors determines both acceptance and rejection of knowledge claims. How the specific choices came about in history can be studied by looking at the concrete instantiations of these factors at token level.

It is mostly thought that accepting the need to move beyond the distinction between internal and external factors in accounting for past science necessarily involves dropping evaluation of past claims to knowledge products in the traditional epistemic sense. Thus there is only an either/or choice to make. Either one accepts a distinction between internal factors and external factors, leading to an evaluative stance with respect to past science, or one adopts a symmetrical position from which a non-evaluative stance follows. To run ahead of things: both main approaches to error will be found to suffer from serious shortcomings. The central challenge that a satisfactory theory of error has to face is to accept the need to move beyond the distinction between internal and external explanations of past science but at the same time to show that this does not exhaust all room for evaluation. In other words, one has to show that it is still possible to work with a context of justification even though it is not possible to specify an internal realm of science any more in unequivocal terms.

Before we deepen the analysis of the two main approaches a number of preliminary remarks is in order. First, the discussion is of a general nature and cannot do justice to the whole variety of approaches to past science that have been formulated. Important differences between scholars that find themselves classified in one major group may therefore go unmentioned. Yet I believe the general approach has important analytical benefits as two very basic attitudes towards errors can be identified and plusses and minuses of these attitudes can be discussed in a meaningful way.

Second, in considering the approaches to past science I do not confine myself to the work of historians only. Work of philosophers and sociologists of science is relevant for the historiography of science in the sense that philosophers and sociologists produce views on how science has developed and how it should (have) develop(ed). It is these views that lead to different emphases on aspects of history and to different kinds of historical interpretations. Although my main interest is in the historiography of science the impact that ideas of other fields have on this profession is important to take into consideration.

Third, it must be made clear in what sense the notion of error is used in this chapter. As is noted by others, 'error' is a multi-faceted phenomenon.[2] Errors occur at many levels and manifest themselves in many ways. Thus in studying errors in past science, historians should not confine themselves only to theoretical error, i.e. the old Baconian sense of 'error as false belief'. Errors in other respects such as in experimental set-ups, in background assumptions, and so forth, need to be accounted for by a theory of error as well.[3] Although most of the discussion that follows focuses on error as false belief it must be kept in mind that this does not cover the whole range of the phenomenon of error. One of the sources of the problems with the two main approaches to error stem from the fact that on both sides parts of the range are left out.

This point can be clarified by making a further distinction that is highly relevant to the study of error. This is the distinction between retrospective error and prospective error. Retrospective errors are errors discovered in hindsight. Claims to knowledge that were held to be true could only be seen as erroneous later on. At the time that these claims were made and defended, errors in them could not be detected, at least not in an obvious way. Prospective error is a notion that pertains to errors that can be foreseen. Examples are a prediction of the likelihood that a hypothesis will turn out to be unsupported by future evidence, mistakes in calculation that could have been avoided, and so forth. The evaluation of prospective errors is the kernel of reliable scientific methodology. In dealing with what we can expect to be wrong we exclude as many errors as we can and arrive at reliable theories. The distinction between what can be foreseen and what cannot be foreseen is relevant for the whole range of error types.[4]

The distinction is so important because it is mainly with respect to retrospective error that a satisfactory theory fails us. Even the most elaborate and also most recent approaches to error, such as Mayo's error statistical approach, for the most part focus on control of prospective error, thereby leaving a vast territory of the phenomenon of error unaddressed. Approaches in the 'error as failures' group often lack the means to engage in retrospective analysis because the historical gaze is confined to very specific local historical contexts. Thus it is not easy to incorporate a retrospective view on errors because this requires one to bring together the uncertainty that confronted people in the past and the more certain points of view that were developed later on in history from which error and truth became visible. How to capture this notion of uncertainty and make it operative as a useful analytical tool with which our understanding of past science can be improved? How to capture the repetitive processes from uncertainty to certainty in science? How to execute such diachronic historiography without inviting the reproach of whiggism, i.e. explaining history backwards instead of forwards?

Historiography of science as a whole is badly in need of a satisfactory theory of qualitative change and this can only be obtained if answers to these difficult questions can be formulated. This chapter contributes to the ultimate aim of formulating a theory of qualitative change in a modest way. It focuses on the identification of the key problems that face the two basic attitudes towards the phenomenon of error. A number of suggestions to move beyond the choice for either the 'errors as obstacles' approach or the 'errors as failures' approach flow from the analysis below. It is these suggestions that at the same time offer the first steps towards answering the questions formulated above.

Errors as Obstacles in the Historiography of Science

The historiography of science started to develop as a scholarly discipline in the decades before the Second World War. Two of the key figures in the early decades of the discipline were George Sarton and Alexandre Koyré. Their views on errors are exemplary for the thoughts on the subject in these decades and are therefore singled out for discussion. First Sarton's treatment of errors will be discussed, Koyré's views on the subject will then be compared to this.

George Sarton has been very much occupied with the legitimatization of the historiography of science as a separate field of study. In a number of programmatic writings he expressed his thoughts on the tasks of the discipline and the need for it in modern society.[5] Sarton saw science as the only human endeavour in which progress had been achieved. However, he recognized that the powerful force of science could be used in the wrong way too. Historiography of science was needed to prevent science from falling into the wrong hands, because only this discipline could perform the following tasks: establish the good tradition that has led to the present-day state of scientific knowledge, show the unity behind the wealth of scientific disciplines by tracing the genesis of all these fields and, finally, show the human aspect in scientific development. He argued that only the historiography of science could function as a bridge between the sciences and the humanities. Without such a bridge, science would fall into the hands of the 'technocrats': specialists in separate fields with no sense of the unity of science and cut off from the tradition in which they work. Such technocrats would invariably overrate their capacities and this hubris would damage society beyond repair.

The humanization of science was one of the most important things for Sarton. The historian was best equipped to demonstrate that science had developed along a non-linear path. Past scientists who struggled but eventually overcame the many difficulties, and in doing so made a significant contribution to the development of man's knowledge, deserved the highest praise. A collection of these achievements made up the good tradition. Evaluation of new ideas could only be done properly with a good understanding of this tradition and should teach present-day scientists above all to be humble. For Sarton, historians of science thus had to act very much as guardians. In modern times no clerical or social hierarchy was available any more to keep people in check. Only historical understanding could prevent things from getting out of hand.[6]

It is evident that the notion of error is important from a Sartonian perspective. Two categories of error are in fact present. The first category consists of superstitions, undeserved privileges, fears, and so forth. This category represents the 'contracting darkness' to which historians should, according to Sarton, pay as much attention to as the spreading of the light.[7] There is, however, a second category of errors too, not part of the contracting darkness. Sarton realized that many knowledge claims that were believed to be true in the past are considered to be wrong nowadays. Still, these erroneous claims could be part of the good tradition of the history of science but only if it could be proved that they constituted important steps forward. Sarton repeatedly emphasized the importance of the scientific method which made such progressive steps possible. Progress in science was thus only possible through continuous criticism of its results: 'There are no dogmas in science only methods; the methods themselves are not perfect but indefinitely perfectible', and therefore: 'Everything is doubtful except the feeling that the margin of error decreases gradually, asymptotically'.[8]

The concept of error is not further developed than this. Errors are seen as obstacles to be overcome and science is seen as a process in which, through the application of the scientific method, erroneous theories are sifted out. Although in Sarton there is no unequivocal negative evaluation of every knowledge claim made in the past that deviates from present-day knowledge, still knowledge claims that we now consider false can only be part of the good tradition if they represented an improvement on previous theories that were even more wrong. Again, the overcoming of this obstacle is the main thing that justifies the verdict of good tradition membership.

Alexandre Koyré's outlook on error was basically the same as Sarton's, but his discussion of errors in past science shows that he had a different interest. Koyré felt that in general historians of science did not dwell on error enough. He acknowledged that what matters from the point of view of posterity is discovery and victory but the difficult processes yielding these victories and discoveries should matter for historians too. Thus Koyré asserted: 'For the historian of scientific thought, at least for the historian-philosopher, failure and error, especially of a Galileo or a Descartes, can sometimes be just as valuable as their successes'.[9] Koyré wanted to grasp past scientific ideas by relating these to the whole mental world of the specific historical periods in which these ideas emerged. From this perspective errors, as well as correct ideas, can be equally valuable for historians because both allow them to retrieve the mental world of the past.[10]

Koyré discusses in detail a captivating case of the simultaneous occurrence of an error in the work of Galileo and Descartes, the latter together with Beeckman. Both arrived independently at the same mistaken formula for speed acceleration during free fall. Initially they both thought that the speed of the moving body was proportional to the distance covered. One of the mistaken consequences that follows from this law is that a body which falls from twice the height that another body falls from acquires twice the speed of that other body. Koyré rules out the possibility that the simultaneous occurrence of this error was the result of pure chance. He thought that errors stand in need of much more explanation than correct claims to knowledge because it is the natural thing to arrive at truth. In case of errors extra reasons must be supplied to explain why they occur. The fact that the error in the formula of free fall came up twice must be explained with reference to impetus physics. This was the dominant way of thinking about motion in Western Europe at the time. It made it hard to conceive of motion in terms of temporal relations.

Only Galileo managed to break out of the old 'thinking cap' and arrived at the correct idea, relating the speed of the moving body as proportional to the time elapsed. The main conceptual change Galileo achieved was to give up a basic idea of impetus physics, namely that the motion of an object was the result of an internal cause of that object. Instead he started to see motion and rest as physical states that could be determined by calculating momentum. Bodies once in motion had no need to stop, or even to slow down, as in impetus physics. Galileo made the important step to conceive of motion as taking place in time and to see distance as a consequence of an essentially temporal reality.[11] Koyré remarks that his efforts can only be appreciated if the mistaken paths are taken into account as well: 'The study of the problem of motion is infinitely instructive – the study of failure always is – and it alone enables us to appreciate and understand the meaning and the importance of the Galilean revolution'.[12]

Koyré saw errors as consequences of the weakness and limitations of the human mind which he attributed to a function of the mind's psychological and even biological conditioning.[13] Like Sarton, he saw errors mainly as obstacles that need to be overcome by the scientific process, the difference being that for Koyré changes in 'thinking cap' were an essential part of this process. Like Sarton, however, Koyré did not analyse the concept of error much further. For this reason we must agree with the conclusion of Hon that even though Koyré attached great importance to the study of the error in the history of science, his own historical studies do not offer much insight in the epistemic phenomenon of error.[14]

Still, it is useful to contrast the way in which historians such as Sarton and Koyré dealt with the notion of error from the more blunt treatment of error that is generally referred to as Whig history. Whig history is a term that covers a variety of meanings. What these have in common is an interpretation of the past in light of the present-day state of affairs. The past is read as a preparation for the present. The contents of past knowledge claims, then, receive epistemic judgements based on present-day knowledge claims. Deviations from these must be seen as errors and the Whig historian can decide to ignore such blind alleys and instead focus on the first formulations of correct theories and how these have come about. Errors serve no purpose in understanding past science. This is not the way Sarton or Koyré, or any other professional historian of science has ever dealt with the phenomenon of error.[15] Sarton and Koyré were certainly presentist in the sense that they interpreted past science in light of subsequent developments, but they very much wanted to understand how errors came about in the past and how they were overcome. This was not done to distribute blame and praise but in order to gain a better understanding of the processes of gaining scientific knowledge. The advancement of science consists in correcting errors and in order to see how this happened the historian needs to look at the way the sources of these errors were removed. It is in this sense they were involved in what Schickore has called 'appraisive history' in which errors are seen as obstacles having a negative impact on science. The errors signify the epistemic distance between past and present and this marks the progress of science.[16]

Errors as Obstacles in Philosophy of Science

Many approaches in philosophy of science share a similar outlook on error as that of the two historians just discussed. Philosophers that have basically seen science as an error-correcting process are, among others, Popper, Lakatos, Laudan and Mayo. Their views will be discussed in that order. In Popper's philosophy the notion of error is central. He famously declared that all the essays and lectures of *Conjectures and Refutations* were 'variations on a single theme – the thesis that we can learn from our mistakes'.[17] In an afterthought on the second edition

Popper added that '*all* our knowledge grows *only* through the correcting of our mistakes'.[18] Popper's theory proposes a method of falsification that unmasks the mistakes. Falsification is basically a procedure of trial and error. Each conjectured theory is subjected to critical tests. If the theory fails to meet one of these, it has to be rejected and replaced by a new one.

Hon argues that we find in Popper the same strange combination as in Koyré: science is essentially seen as a process of error elimination but a theory of error is not developed. Popper nowhere investigates the conditions of error, the various kinds of error nor the variety of effects errors can have.[19] Two reasons for this can be given. First, Popper neglected the context of discovery and it is in this context that errors come about in the first place. Second, Popper thought his logico-deductive framework and the empirical testing needed for the critical tests were both relatively straightforward. However, both empirical testing and rational falsification can be profound sources of error. But in order to grasp this, one needs to study particular aspects of historical contexts and that was not something Popper was very much interested in. In later years Popper actually made an extra point of the contextual blindness of his evaluative procedures. He aligned his view of the development of science with evolutionary epistemology and it is the analogy to the blindness of natural selection that Popper wanted to secure.[20]

In a significantly more subtle way, we can find the same attitude as Popper in the work of Lakatos.[21] Lakatos made a distinction between internal and external history. Internal history is that part of the history of science that can be rationally reconstructed; external historiography should supplement these rational reconstructions with explanations whenever the historical course actually taken deviates from the rational model. For Lakatos historiography depends on the selected model of rationality. This model is the embodiment of what the philosopher (or historian) thinks is the correct way of doing science. Models of rationality can be quite different, such as Popper's falsification procedure, various forms of inductivism, various forms of conventionalism or Lakatos's own methodology of scientific research programmes. Anything in the history of science that falls out of the selected rational model must be empirically explained by external historiography, that is, by social or psychological factors. The boundary depends on the preferred model of rationality.

Lakatos is a bit ambiguous about the place of errors in his own model of rational reconstruction of research programmes. In some places he opposes rationality and errors outright: 'One can or should not explain all history of science as rational: even the greatest scientists make false steps and fail in their judgment'.[22] But in other places he also suggests that the problem of deviation can be solved internally. His methodology of research programmes does allow for temporal recessions in the development of the programme as a part of internal history. Errors can, then, also be part of that internal history but only as an empirical explanation for a recession in the development of a research programme.[23] Even though upon this reading errors can now be part of the rational reconstruction of past science, to regard errors as obstacles to progress is still the dominant view. Errors are regarded as impediments to the progress of science and it makes no principled difference whether they are the result of social, psychological, religious, ideological, biological or conceptual conditioning.

Larry Laudan has taken the Lakatos methodology of research programmes one step further. He distinguished between the *general* nature of rationality and *specific* forms of rationality set by context-specific parameters. The general nature of rationality holds in all periods. It consists of certain attitudes that are always present: for example to accept the most effective problem solvers. What these problems are and what are seen as effective solutions can then still be context dependent. Specific rationality is further defined in terms of the availability of reasonable choice options in a given time and place. In Laudan's model of science it becomes very important to zoom in on historical contexts and see what kind of rationality was operative. Contrary to Lakatos it is not needed to settle on one type of rationality only in order to carry out rational reconstruction of the past. Instead Laudan says that 'We must cast our nets of appraisal sufficiently widely that we include all the cognitively relevant factors which were actually present in the historical situation.'[24] It is with respect to these factors that one has to assess whether new theories are progressive because it is with respect to these factors that problems are formulated, their importance weighed and solutions to them accepted or rejected.

According to Laudan we have no way to ascertain whether rational procedures lead to truth, but we can establish whether rational procedures, and even changes in rational methods, lead to progress. The measure of progress in science for Laudan is the increase in problem-solving capacity of theories or research programmes in comparison to alternatives. Thus he disconnects rationality from truth but a close connection between rationality and the notion of progress is maintained.[25] This does not lead to an 'anything goes' attitude because the demands on the rational standards in any period are quite high and because of the comparability in problem-solving capacity of theories over time. This comparability stretches further than specific localities and can be extended to our own scientific claims. The focus on problem solving as a measure of progress is then flexible enough to allow for both contextual assessments as well as translocal assessments of knowledge claims.[26]

Since Laudan disconnects truth from rationality he can also not speak of errors in science in terms of truth. He can only speak of errors in terms of an increase or a decline of problem-solving capacity. It is, however, not a straightforward task to assess progress in terms of problem-solving capacity because typically there is more than one problem at stake. The weighing of problems is thus very important. As Laudan wrote in a later work: 'Indeed, on this model, it is possible that a change from an empirically well-supported theory to a less well-supported one could be progressive, provided that the latter resolved significant conceptual difficulties confronting the former'.[27]

Things only go wrong in Laudan's model when the general or the specific concepts of rationality are violated. It is in such cases we must appeal to a sociology of knowledge in order to find explanations for this behaviour. For Laudan, the sociology of knowledge can contribute to the understanding of science in three ways. First, it should explain irrational behaviour such as the acceptance of a research tradition that is less adequate than an existing rival, the pursuit of a non-progressive theory and the attachment of the wrong weight to a problem. Second, it must give the deciding factors when a choice has to be made between theories or research programmes that are equal in their problem-solving capacity. Third, it should investigate the social structures which make it possible for science to function rationally in the first place.[28]

We can now see that even in the highly context-sensitive framework of Laudan there is still an asymmetrical explanation of scientific claims. Deviations from his model of rationality need to be accounted for by social factors; correct actions can be rationally explained. This approach may be called a sociology of error. We are drawn to the conclusion that all approaches to science that see errors primarily as obstacles to progress do also maintain some form of asymmetry between rational and social explanations of science. Thus they all rely on some form of a sociology of error.

The last approach to error in our discussion that can be headed under the 'errors as obstacles' approach is Mayo's programme of error statistics. Mayo is the philosopher of science who has perhaps paid the most attention to the phenomenon of error in recent years. As a consequence, she has developed an elaborate theory of error. According to Mayo, in many philosophies of science

> Little is said about what the different types of errors are, what specifically is learned when an error is recognized, how we locate precisely what is at fault, how our ability to detect and correct errors grows, and how this growth is related to the growth of scientific knowledge.[29]

We have indeed seen that in the work of Koyré and Popper such questions were hardly posed. Her aim is to present a theory of learning from error that is sensitive to all these issues.

Mayo signals a blind spot in Popper's model with respect to learning in science. The only ways of learning Popper's falsification procedure yields are either by rejecting hypotheses that are falsified or by corroboration of hypotheses if a test does not falsify them. Mayo argues that on the one hand falsification destroys theories too drastically and on the other hand corroboration offers no more than weak support of hypotheses. Popper's procedure does, therefore, never yield reliable knowledge. In her own model she wants to secure reliable knowledge by constructing 'arguments from error'. In short, this says that knowledge is reliable when we can be almost certain that errors are absent.

What do these 'arguments from error' look like? First, it is important to grasp Mayo's starting point that a degree of indeterminacy must be accepted in modern science. No perfect control of experimental findings is possible and hence we must resort to probabilistic reasoning of some sort. This is the (only) way to deal reliably with experimental outcomes which invariably contain a degree of indeterminism. Mayo does not opt for a Bayesian approach because she finds this approach unsatisfactory. Instead she prefers a statistical approach that is derived from Neyman and Pearson. Central in this approach is the idea of the severe test. A severe test is a test procedure that has a very low probability of letting a hypothesis pass when this hypothesis is actually false. In contrast with the Bayesian approach to probability Mayo does not attach error probabilities to the degree of belief but probabilities are attached to the *methods* of inquiry. These are thought to be properties of testing procedures and *not* of theories themselves. The error probability, then, is the frequency of trials in which a test passes a hypothesis given that the hypothesis is false.

This way of handling probability yields the following argument from error: an error is most probably absent when a procedure of inquiry that has a high probability of detecting an error detects no error and/or an error is present when a procedure of inquiry that has a very high probability of not detecting an error nevertheless detects one.[30] One learns in science not by confirming a hypothesis by evidence but by excluding discordant information. Although different from Popper in many respects the model still has an air of the falsification procedure, and is sometimes called Neo-Popperian.

Mayo argues that a lot of present-day scientific research can be a captured by this formal approach. However, she does recognize that an informal inquiry into errors is needed too. Checks must be performed by testing for canonical mistakes. Examples of such mistakes are: mistakes about the quantities or value of a parameter, mistakenly seeing experimental artefacts for real effects, mistakes about causal factors and mistakes about assumptions of experimental data. These involve qualitative judgements and not just quantitative analysis. Mayo accounts for such canonical mistakes with the notion of error repertoire. Each scientific discipline has built a set of procedures that exclude making canonical mistakes. In spite of the diversity of errors that can occur, the argument from error remains the same: severe testing procedures (whether statistical or taken from the error repertoires) ensure a high probability of absence of errors and hence the reliability of knowledge claims.

The reliability is also ensured by the demand that primary hypotheses, data models and experimental models must be probed for error independently. It is this demand that must help to overcome the Duhem problem. This is in line with the New Experimentalist focus on experiments as a way to avoid common philosophical problems:

> New Experimentalists share the view that a number of problems, such as the under-determination of theory by empirical knowledge, the theory-ladenness of observation, and extreme sceptical positions, stem from the theory-dominated perspective on science. They defend that focusing on aspects of experiments and instruments in scientific practice holds the key to avoiding these problems. Some of the key figures of this movement in the 1980s and early 90s are Ian Hacking, Nancy Cartwright, Allan Franklin, Peter Galison, Ronald Giere, Robert Ackermann, and more recently, Deborah Mayo.[31]

The main claim of this movement is that experiments can be set apart from theoretical discourse, i.e. experimental results can be processed in more than one theoretical framework. As Chalmers put it: 'The New Experimentalists are generally concerned to capture a domain of experimental knowledge that can be reliably established independent of high-level theory'.[32]

Within this context Mayo focuses on experimental testing procedures which she interprets as paradigm-independent methods of research. According to her, a steady increase in frames of reference in philosophy of science to capture the problem of theory ladenness is an ill-guided approach:

> The response to Popper's problems, which of course are not just Popper's, has generally been to 'go bigger', to view theory testing in terms of larger units – whole paradigms, research programs, and a variety of holisms. What I have just proposed instead is that the lesson from Popper's problems is to go not bigger but smaller.[33]

Every move to go bigger yields the problem that a demarcation between scientific knowledge claims and non-scientific knowledge claims is increasingly harder to draw. 'To go smaller' means focusing on experiments and here method is more important than theory. As Mayo puts it: 'What we rely on, I will urge, are not so much scientific theories but *methods* for producing experimental effects'.[34] Methodological rules for experimental learning are strategies that enable learning from common types of experimental mistakes. Upon confrontation with errors that are not yet established, Mayo advises strengthening the 'error repertoire' of the scientific discipline in question. As our methods of error control grow our knowledge claims become increasingly reliable.

To see methods in science as the embodiment of overcoming errors made in the past is a very interesting thought. Allchin even suggests that philosophers have more or less neglected to study science from the perspective of error because the errors are dissolved in these methods and have become invisible. An interesting research avenue for historians, then, is to open up these methodological 'black boxes' in order to improve our understanding of the role past mistakes have played in the development of the respective scientific disciplines.[35]

There are many parallels between Laudan's philosophy of science and Mayo's approach to error. For Laudan, the chief element of continuity is the base of empirical problems. This means that the problem-solving capacity of these theories can be compared and assessed. Mayo also speaks about problem solving in terms of weighing of empirical problems. The most important problems need to be solved first and, like Laudan, she contends that this might result over all in a decrease in the number of problems solved. We can clearly hear an echo of the voice of Laudan in the following citation: 'Knowledge of relative weight and relative number of problems can allow us to specify those circumstances under which the growth of knowledge can be progressive even when we lose the capacity to solve certain problems.'[36] Laudan's rationality concept is embodied in Mayo's work in methods of detecting errors in science. Where in Laudan's model standards of rationality change from time to time, in Mayo's model error repertoires change. It is improvement in standards of rationality and error repertoires that ensures reliability of knowledge claims because these improvements lead to the exclusion of a greater number of errors.

For Mayo there is just normal science, understood as standard testing.[37] This paradigm independent testing procedure is aimed at exclusion of errors. This is seen as the motor of science and science is thus basically seen as an error-correcting process. We must conclude that Mayo's model of science therefore also falls within the group that treats errors as obstacles to progress. Her close alignment to Laudan supports this conclusion. The fact that the rationality of science is embodied through probing for errors via severe testing and execution of error repertoires that are paradigm independent supports this conclusion as well, because all the approaches to past science in the 'errors as obstacles' group grant rationality a special status. Although Mayo does not discuss the topic it can be inferred that when reasons must be sought why in history procedures of severe testing for error were *not* followed, some sort of social account will figure as an explanation.

Errors as Failures

Robert K. Merton, who can be said to have inaugurated the social study of knowledge, asserted that 'The sociology of knowledge came into being with the signal hypothesis that even truths were to be held socially accountable, were to be related to the historical society in which they emerged.'[38] However, it was only with the advent of the sociology of scientific knowledge (SSK) in the 1970s that the full implications of this idea came about. The SSK School started to reject asymmetrical explanations of errors and truth. The idea that right actions carry their own motives and only wrong actions stand in need of explanation was found unwarranted. David Bloor opposed a sociology of error in which nothing makes people do things that are correct and all incorrect actions need to be socially accounted for.[39] For proponents of SSK each form of rationality must be socially accounted for. As Bloor puts it: 'There are no limitations which lie in the absolute or transcendent character of scientific knowledge itself, or in the special nature of rationality, validity, truth or objectivity'.[40] Thus the distinction between external (or social) factors and internal (or rational) factors simply disappears.

Bloor was the first to introduce a principle of symmetry in the science studies.[41] This principle orders the science student to treat all knowledge claims symmetrically. For Bloor the same *type* of factors determine both acceptance and rejection of claims to knowledge. While other factors are relevant to the scientific process the determining type of factors is categorically social. A scientist can do research individually, come up with experimental results and theorize about these. But when it comes to sustaining such knowledge claims he has to enter the social sphere and often engage in a debate with other people. Such debates are invariably settled along the lines of social behaviour which are best studied by sociologists. In speaking of social factors Bloor thought both about 'micro' processes of negotiation, such as relations of authority and trust but also about 'macro' processes such as the working of political systems.

The principle of symmetry leads to a different view on the very nature of what errors are. It leads to the dispersion of the distinction between true and false beliefs in the traditional epistemic way. Knowledge is not defined as true belief but as authorized belief. Knowledge formation is a result of negotiation and in this process meanings are created. Ideas no longer have explanatory value in SSK. Ideas are the things that *need* to be explained and are interpreted only by assigning them a function in a given society. For Bloor, negotiations have a generative character and the meanings created have relevance to specific contexts only. This position is generally called finitism, another apt term is functionalism. Social factors have relevance with respect to the functioning of a particular society, these functions give the social factors their determinative character. Knowledge production, then, is seen in relation to the functioning of a specific society.

Looking at the function of knowledge, and at the same time at the concepts of truth and error, requires one to zoom in on specific historical contexts. The sociologist/historian is particularly qualified to investigate the functionality of (past) societies. With respect to such frameworks, scientific controversies can be studied. The scientific controversy brings about conflicts of interests. People are always engaged in pursuing their interests. The knowledge claims of winners of controversies have gained acceptance because the winners were better attuned to the demands of society or because they were simply more powerful. There is still a notion of error present in SSK but it is interpreted only in terms of failure. One can only fail with respect to a given design with specified functions. The knowledge claims the 'losers' in these conflicts made are not seen as wrong in the traditional sense but only as failed attempts to gain acceptance. As Schickore formulates it, errors are what the community decides they are.[42] They are always the result of a social decision; a decision is never purely based on epistemological content, whatever the status of this content.

The view on errors as failures and the interpretation of justified and unjustified knowledge claims as results of social negotiation processes has had a tremendous impact on the historiography of science.[43] Many studies of error in terms of failure have been produced.[44] The study of technological failure, for example a plane crash, is sometimes used as a template in order to arrive at a detailed account of why things went wrong in 'theoretical' science.[45] It is important to recognize that one can only make such analysis of past science if it is assumed beforehand that a substantial part of the functional design of the socio-cultural context of which the scientific failure being studied was a part can be specified. As in some of the approaches from the 'errors as obstacles' perspective all attention in the 'errors as failures' perspective appears to go to prospective error again, although the analysis runs in completely different terms.

The SSK approach has been further extended in the last two decades with the so-called 'posthuman' approaches to science.[46] An example is Latour's 'Actor Network Theory' which focuses on the networks in which knowledge claims are sustained.[47] Latour's clearest example is perhaps found in the *Pasteurization of France* in which he demonstrates how Pasteur's knowledge claims, for example with respect to microbes, got accepted by looking at Pasteur's network-building abilities. In many ways, Latour's approach goes beyond the social turn of SSK, but negotiation and the settlement of conflicts remain at the core of his network theories. Thus knowledge claims that do not find acceptance are the result of failed attempts at network building. This theoretical model is not so new but has gained increasing attention in the last couple of years among historians.[48] The field is undergoing a 'spatial turn' which can be witnessed by the many 'circulation of knowledge' research projects that have been initiated. It is Latour's 'Actor Network' model that often offers the conceptual resources for such studies.

Constructivist thinking about knowledge has also led to a practical orientation in the historiography of science. A good example is the work of Pamela H. Smith.[49] She equates the ability to make things with having scientific knowledge. It follows that the norm in science is embodied cognition, i.e. the practical and material circumstances scientists work in and the social entrenchment of these working places. The norm is not texts and the content of these, although the writing of texts can itself be seen as a craft and studied in a similar way as the other scientific practices and the skills needed to perform them. This insistence on craftsmanship is important since Smith thinks the boundary between pure science and applied science needs to be broken down. For her there is no pure science, science always has an applied nature. Smith and her group have actually re-enacted past scientific practices in which they tried to stay as close as possible to the materials and devices past practitioners had to work with, for example re-enacting the work of seventeenth- and eighteenth-century goldsmiths. They discovered that experimenting is very difficult and that it goes wrong on many occasions. Succeeding, in other words, is a process of trial and error.[50] Only after many pitfalls did people in the workshop obtain the desired results. In the process skills are developed in order to avoid failures in the future.

This type of historiography can also interpret errors in light of clearly specified functions provided by the historical context in which the errors occur. Science is seen as a means to reach an end. We are led to the conclusion that both the social or contextualist turn, the spatial turn and the practical turn in the historiography of science all consider the phenomenon of error in terms of failure. Moreover these analyses are mostly confined to prospective error only. In the next section it will be argued that an important aspect of dealing with errors in past science is left untouched in this way, namely that the source of error is very often the *absence* of clear functional designs and can only be established retrospectively.

Shortcomings of the 'Errors as Obstacles' Approach

Two very distinct ways to deal with errors in past science have been recognized in the previous sections and the main ideas behind these approaches to error have been displayed. In this and the next section both main approaches to error will be critically reviewed. It will be shown that both of them suffer from serious shortcomings and hence both are unsatisfactory for the study of the phenomenon of error in past science.

The 'errors as obstacles' approach is unsatisfactory for three reasons. First, it is hard to maintain a strict demarcation between rational explanations and social explanations of past science. Too much evidence in the historiography and sociology of science has accumulated that signals that in theory choice it was never one factor that was decisive but that a combination of factors was always at play. On the philosophical level, granting the factor of rationality a special place requires one to answer arguments of incommensurability, under-determination, theory ladenness of observation, and so forth, which has proven to be notoriously difficult. Maintaining the strict distinction, then, is not historically adequate and invites more philosophical problems than it solves.

The second problem with the conception of errors as obstacles is that it is hard to account for positive effects of errors in this way. As a matter of fact, if the prescriptions of some philosophers were followed all the time, past erroneous ideas should have been eliminated at a much earlier date than actually happened. The positive effect of errors would then have been thrown out with the bathwater as well. However, a number of scholars has pointed out that we learn from error in subtle ways and that positive epistemic roles can also be ascribed to errors.[51]

One way in which errors can play a positive role is when a mistaken argument or proof leads the way and reaches a useful idea that would otherwise perhaps not have been obtained. Such a discovery can then be said to be right for the wrong reasons. A number of Einstein's mistakes can be interpreted as such.[52] Closely related is the opening of a new perspective on things that turns out later to have been wrong. Still, this new vista may constitute a leap forward because it destroys old ways of thinking and opens up new possibilities for research. The Cartesian corpuscular worldview is now thought to be incorrect, but it helped to overthrow Aristotelian doctrine and made a significant step towards perceiving the world through a quantitative instead of a qualitative lens.[53]

A third positive role of errors can come about through the mirror effect: if you know what is wrong it is possible to see what is right by contrast. This effect is captured in catchy phrases such as 'turning is learning' or *contrariorum eadem est scientia*. We never know what a thing really is unless we are also able to give a sufficient account of its opposite. Thus we may shed light on the concept of beauty by studying its opposites. Mistakes/errors in science can in the same way illuminate what real knowledge is.[54]

Even if the ability to specify errors fully does not lead to knowledge, they may yield valuable information which may be used in further theorizing. This is the fourth way in which errors can be said to have a positive epistemic role in the development of science. Wimsatt has indicated six ways in which this can come about by looking at the role models play in science.[55] Wrong models may serve as a starting point for a series of more complex models, suggest new tests or refinement of established models, serve as templates that account for large-scale effects and that make smaller effects noticeable, serve as limiting cases that are true under certain conditions and define extreme cases between which other cases lie. Finally, wrong models may provide a simple arena for determining some properties of a system. Thus in general false models give rise to anomalous results that stimulate the development of more sophisticated models and theories.

All this is not to say that all errors must be seen as fertile. Quite the contrary, errors can often be much less useful. Perhaps it is a good idea to distinguish between 'good wrong' and 'bad wrong', where 'good wrong' is in one way or another conducive to eventual knowledge. Fertile errors are then 'good wrong' and obstacles to progress are 'bad wrong'. In this sense perceiving errors as obstacles can still be a relevant perspective.[56]

The third reason why the 'errors as obstacles' approach as a whole fails to be satisfactory is that insufficient attention is paid to the historical situations in which errors came about, are detected, are corrected, and so forth.[57] On the one hand this involves a neglect of the context of discovery but on the other hand it has effects on the consideration of the context of justification as well.

If we look at the discovery context first, we see that philosophers such as Popper and Mayo hardly pay attention to the emergence of theories or even of testing procedures. The work of Mayo, for example, does not show much historical interest and is intended to aid present-day scientists in arriving at the best founded knowledge claims possible. As the present interest lies in the study of past science, questions emerge such as: where do the error repertoires come from? Where do assessments of probability of testing procedures come from? How and why do changes in probability estimates and in error repertoires occur over time? Do scientists follow Mayo's prescriptions and if not, what does that say about the knowledge claims scientists produce?

To be fair, Mayo (this volume, p. 65) recognizes the problem as she writes: 'What about the work that goes into designing (or specifying) hypotheses to test or infer? Much less has been said about this, and it is a central gap to which I encourage philosophers of experiment to fill.'[58] Yet she also resorts to the classical 'trick' in the philosophy of science of making justification the business of philosophers and discovery an area of contingent historical practice which does not present itself as a fruitful area of research for philosophers because it is not systematic (this volume, p. 67). Because in Mayo's framework it is the methods of research that ensures reliability of knowledge claims and because these methods are constantly *improved*, this is not a satisfying answer. Moreover, there *are* systematic things to say about the discovery process and this is indeed an area in which much more work can be done.[59]

But in the context of justification, problems arise as well. In the 'errors as obstacles' approach one method of testing hypotheses is defended and it is this method of testing that sifts out errors. However, this is often about prospective error: the number of errors that are excluded is limited by the mechanism that is designed to sift these errors out. No procedure of known errors can exclude all of them beforehand. There may always be sources of error that are overlooked, hence these have to come about in other ways. If existing procedures are not able to capture errors they can only be updated *a posteriori* and therefore we need a theory of retrospective error as well.[60]

Critical testing by rational procedures is not the only way in which hypotheses are confronted in science. One can also test theories in other ways. The most important is by confrontation with alternative hypotheses that offer a new perspective on a problem. Quite often only *after* such a thing happens errors in the previous perspective become visible. It is precisely in this sense that Hasok Chang has expressed the feeling that Kuhn's ideas of paradigm shifts get flattened too much in Mayo's model.[61] The challenge of course is to make such shifts in perspective part of normal science without running into the problem of incommensurability. If this can be achieved, as I think it should be, then normal science is not just standard testing as Mayo asserts.[62] There is more to it and this involves a more complete treatment of the phenomenon of error.

The main drawback of Mayo's treatment of error is that no view is developed on the diverse ways in which conceptual frameworks interact with experiential practice. As long as errors are sufficiently understood Mayo's model works, but what if this is not the case? How to severely probe for error if a clear conception of what one is testing or what can be expected from experiments is not present?[63] The discussion of the third argument against the 'errors as obstacles' approach has focused on Mayo's error statistical framework (mainly because it is the most recent and most elaborate theory of error around) but the conclusion holds for the whole 'errors as obstacles' group of approaches. A neglect of the role of the unforeseen is present in all the 'errors as obstacles' approaches that can be found in the history and philosophy of science. Prospective error is mostly focused upon and this leaves all kinds of interesting issues pertaining to the phenomenon of error untouched.

Shortcomings of the 'Errors as Failures' Approach

Philosophical models that do not square enough with historical practice lose a lot of their explanatory value. The view on errors as failure or the consequence of negotiation must be seen as a reaction to the dominant 'errors as obstacles' trend which historians increasingly found too idealistic. These reactions came about strongly in the 1970s. In this period the expression 'whig history' in relation to the historiography of science, which is now commonplace, came into frequent use.[64] This indicates that a need was felt to draw a sharp distinction between what historians were doing at the time and what historians were doing before.

A number of just criticisms against this earlier historiography were involved. Whiggish sins such as studying the past as a preparation for the present, making unfair judgements by failing to evaluate past practitioners in terms of their own milieu, uncritical acceptance of scientific controversies by the winners of these controversies, unjust bias towards persons, countries, or even historical periods and so forth, must be avoided. For all these reasons it was thought that any presentist attitude leads to highly selective forms of historiography and, eventually, misrepresentations of the past. Another problem with Whig history is that such representations need constant revision because the present constantly changes. This indeterminacy makes historiography the toy ball of present-day developments, which is obviously unsatisfactory.[65]

While all these criticisms are, in themselves, just, they have also led to an exaggerated contextualism and oversensitive attitude to all forms of presentism or progressivism. As one historian noted: 'The reproach has become a shibboleth, which is placed in position against anyone who claims that science is in progress at all'.[66] Indeed, non-evaluative historiography has become dominant in the profession and historians are inclined to study errors as failures eschewing traditional epistemological issues. Hon, Schickore and Steinle agree with this and take the fear of whiggishness as the most important reason holding historians of science back from a more natural treatment of errors in past science.[67]

But why do we need to mourn this? Is there not a perfectly legitimate way available to deal with the phenomenon, i.e. interpreting errors solely in terms of failure? A number of reasons can be given why this way of approaching errors is not satisfactory either. First, humans stumble and make mistakes all the time. Often it is said that it is precisely this that makes people 'normal'. Persons who do not make mistakes are seen as super-humans or saints. Now approaches to past science based on some sort of symmetry principle stress above all that science is a man-made fabric. Therefore it is strange that one of the basic human character traits is explained away in symmetrical analyses of past science. It is highly doubtful that without some sort of assessment of knowledge products the past can be aptly understood. As Sargent put it: 'Because of the inherent fallibility of human senses and reasoning, such a qualitative check is absolutely necessary'.[68]

Scientists themselves have been well aware of their limited capabilities. They know they are likely to be wrong, that they are prone to fall into error and bound to fail at many occasions. The 'pessimistic meta-induction' and the 'no privilege argument' are not news to them. Albert Einstein, for example, scolded himself for introducing the cosmic constant. He thought it was bad science to introduce constants without reference to a physical cause. He regarded the cosmic constant as the biggest mistake of his life.[69] Max Weber asked in his lecture 'Wissenschaft als Beruf' why people are willing to choose science as a profession if they can be certain that their contributions will be refuted in the future: 'Jeder von uns dagegen in der Wissenschaft weiss, dass das, was er gearbeitet hat, in 10, 20, 50 Jahren veraltet ist ... jede wissenschaftliche "Erfüllung" bedeutet neue "Fragen" und *will* "überboten" werden und veralten'.[70] Luckily Weber was not altogether pessimistic about science but nonetheless his point is clear: scientific theories are often corrected, adjusted, refuted and so forth, and practising scientists have to live with this.

From a functionalist perspective, however, it is very hard to account for *qualitative* changes in science. First, the analytical tools are designed to gain an understanding of local contexts only. Second, questions about how different contexts relate to each other or how shifts and transformations occur in them can only be explained in the same terms. As a consequence, an analysis of past science cannot extend the direct causal connections between a context and the one it is related to. Moreover, changes can only be accounted for in causal or quantitative terms. A change in function, for example, can be explained by an account that demonstrates why a given function ceases to be useful in a given context. Perhaps it is replaced by another function meeting a new demand. Whether such changes must be seen as improvements is however not a question that can be addressed.

The decrease in scope of research may be a result of desiring a different type of historiography. But the change in approach that followed has made evaluative analyses of past science almost impossible to execute. The challenge to regain some qualitative analytical grip is to find an appropriate diachronic widening of scope. Again, just as in the 'errors as obstacles' approach, but in a different way, it is only prospective error that can be captured in the 'errors as failures' approach. Epistemic uncertainty is recognized in the 'errors as failures' approach but the explanation of the way uncertainty is resolved requires certainty in terms of the socio-cultural factors that were at play in historical contexts. Retrospective error requires one to take some uncertainty about socio-cultural factors in historical contexts into account as well. In the 'errors as failures' approach it is difficult, if not impossible, to realize this.

The required neutral attitude to all past scientific contributions worsens the situation because it effectively downplays the role of the historian to a mere onlooker of past events. This leads to a further, and perhaps surprising drawback of the 'errors as failures' approach, namely that it hampers the aim to reach the very goal of this approach, which is the understanding of the local historical contexts. First, standards of error have changed in the past and the views on error that a past practitioner had can be the key to understanding his total approach to science.[71] Second, understanding the past may be helped in certain cases if we assume that certain phenomena scientists had to deal with are very much comparable to phenomena scientists have to deal with now. Improved present-day understanding of these phenomena can then possibly help to understand the struggling of past scientists and quarrels they had with each other.[72] Both these clues to gain understanding of a historical context are not available to the social contextualist because they require one to 'zoom out' of local context and make diachronic comparisons.

To sum up, the 'errors as failures' approach has a number of shortcomings that mainly flow from the narrow contextual focus on past science. This makes it hard to account for qualitative change and hence errors are interpreted in the failure sense only. But this way of dealing with error requires clarity about all socio-cultural factors at play in a given historical context. The type of error addressed in such functionalist terms can thus only be prospective. Again, interesting historical questions about theory change cannot be answered satisfactorily because only part of the phenomenon of error is addressed.

Conclusion and Ideas to Move beyond the Dichotomy

We have found that the two main approaches to the phenomenon of error in the historiography of science do not yield a satisfactory treatment of errors because both suffer from huge shortcomings. Treatment of errors as obstacles leads to historical accounts that need to use a demarcation between rational explanations and social explanations of knowledge claims. With enough sophistication this does not have to lead any more to passing unfair judgements on historical actors.

However, in light of all the research after the social turn in science studies, a strict demarcation between social and rational factors is hard to maintain. Further, the 'errors as obstacles' approach does not capture the whole game of science, only a part of it. Insufficient attention is paid to the frameworks in which errors emerge and are discovered. Moreover, *positive* effects of errors are very hard to account for, if noticed at all.

Seeing errors solely as failures or as a result of negotiation addressed a number of these problems but turned out to be equally problematic. This approach withdraws epistemological concerns completely from the historiography of science.[73] With this approach it is not possible to account for change in qualitative terms while the fallibility of humans cries out for such evaluations. As a matter of fact no proper theory of learning can be built on this approach to error at all. The functionalist outlook also has great difficulty accounting for situations in which it is felt that something is not in order but this something cannot be clearly specified yet. Even if the aim of these approaches is only to achieve accurate descriptions of past historical events and episodes the non-evaluative stance makes it fall short in this respect too. Access to continuity of phenomena and to standards of error actually in use by past practitioners as tools of research is not available. The proliferation of the symmetry principle has led to the erasing of more and more boundaries. But this has also meant giving up selective criteria with which the history of science may be approached. Everything that happened may become relevant to the study of past knowledge claims. Historiography of science is then led to producing histories of everything, which effectively means histories of nothing.[74]

Clearly something is wrong in the historiography of science, as the field lacks a satisfactory approach to the notion of error. However, the discussion has indicated all the central points on which a search for a satisfactory theory of error must focus.[75] These involve a (re)consideration of the notions of uncertainty, rationality and diachronicity as well as developing a sophisticated form of presentism. It can only be done briefly here but I would like to end the chapter with a few comments on all of these which might help to set the agenda for future research.

First is the notion of uncertainty. We have seen that Mayo developed her approach to error in light of uncertainty at the frontier of science. She attempts to cover this in terms of indeterminacy. Her solution to gain as much control as possible on situations where evidence (and theories) are indeterminate via probabilistic reasoning, is not unlike other solutions of philosophers. Her innovation is to attach the probability that errors are present to methods instead of theories. However, although this moves away from theory and as such avoids a number of philosophical hard nuts, a move 'further away' is perhaps needed. Uncertainty properly understood is first and foremost an aspect of persons. Realizing this leads to a change in perspective on past science. For example, it provides more space to consider uncertainty in historical situations in which no prospective calculation of errors is possible. The change in perspective may offer new analytical ways to account for retrospective error in the history of science, a problem for both main approaches to error, as discussed in the previous sections.

Interestingly a number of recent studies have started to explore this way of dealing with uncertainty. Nowotny, Scott and Gibbon have argued that we do indeed lack absolute standards of evaluation.[76] But the conclusion of this must not be to drop every evaluative stance. On the contrary, because of the lack of absolute standards a constant criticism of scientific results is needed. Hence openness for debate and confrontation with other views must at all times be strongly encouraged. We have seen that it is precisely this that is often lacking in the 'errors as obstacles' approach because in this approach confrontation of hypotheses is only done via *a priori* specified rational procedures.

A promising new analytical concept that relates to this discussion is the notion of 'going amiss'.[77] The notion of 'going amiss' pertains to situations in the past in which people felt something was not right but they were not in a position to tell what it is. Only after time elapsed, new discoveries were made and so forth, it became possible to indicate where the errors in the past situation should be located. The 'going amiss' notion allows us to treat every past participant in a scientific debate with the utmost respect because even though his or her contribution to this debate may have been false on later grounds, in the time itself it was respectable enough to defend. Yet the notion does not force us to be neutral on all contributions to past science. It shifts the general focus of research to the question how uncertainty was gradually resolved in a given historical episode. This can be compared to the development of other cases of theory development in science and perhaps consequently generalizations about the scientific process are possible on these studies. The only thing that has to be assumed by the present-day historian is some form of agreement between past and present on how people react in typical situations, for example the assumption that when scientists are confronted with uncertainty they attempt to resolve it. Science, then, should not be seen as a process of removing obstacles to progress but as a process of removing uncertainty. Although removing obstacles to progress can still be part of the business of science, the suggested change in perspective makes it no longer the primary concern. It also opens the door to consider a variety of ways in which the uncertainty in the past could have been overcome, which does not only have to be through the application of a standard rational procedure.

Another desideratum that stands out from the discussion is the need to find a right diachronic zoom to study past science. The pressing question is what we can take as our platform in the present to study the past without running the risk of being accused of presentism, whiggishness and so forth. Hasok Chang, too, has pointed towards this issue in a recent article:

I believe that history of science as an academic discipline is by now strong enough to throw off the crutches of neutrality with respect to the sciences; what we need instead is a declaration of independence. Accepting (and moving beyond) the inevitable present centeredness of historiography, we can have a self-confident conception of historiography founded on the recognition that the historian is a free agent, despite her/his obvious rootedness in present society and science. The question is: which part and which version of the present do we choose to take as our platform? And as usual, freedom comes coupled with responsibility. We historians need to face up to the implications and consequences of the judgements we do and must make. I am much happier to accept that burden of responsibility, than to hide behind a murky notion of neutrality.[78]

Chang, however, has not attempted to indicate what the elements of this platform should be. Suggestions that spring to mind are an assumption of continuity of phenomena, assumptions of agreement of cognitive attitudes of past practitioners and present-day historians, usage of constitutive anachronisms and a newly interpreted notion of rationality (see below, p. 37).[79] Another idea to achieve a right diachronic zoom on the past is to consider scientific change in the context of disciplines, but only if the latter are interpreted in a dynamic sense.[80]

Both the ideas of seeing resolving uncertainty as the primary driving force in science and setting up a platform containing resources with which the historian can approach past science are important steps towards maintaining an evaluative context while giving up on asymmetrical analyses of past claims to knowledge. Speaking of demarcation, the above discussion has also clearly shown that the central notion in the whole discussion on error is the notion of rationality. The two main approaches leave us no other choice than either to accept that the past is frozen into present-day rational procedures or accept that the past is frozen into distinct localities, each with their own set of rational procedures. Neither of these are satisfactory.

Something strange is going on here as in *both* approaches strict ties between rationality and progress are maintained. In the 'errors as obstacles' approach, rationality is the sole vehicle of progress. The concept provides the means to overcome errors and this then leads to progress. In the 'errors as failures' approach no such context-independent concept of rationality is available any more and hence all talk of progress is dropped. You either have both of them or you have neither, which shows that the assumption that the two are strongly connected is shared by both camps. Laudan has argued that next to this there are a number of other common assumptions between what he calls positivist and post-positivist approaches to past science.[81] In effect this means that the post-positivists have left some of the positivists' assumptions about science intact. They have just drawn different conclusions from the same premises. Laudan argues perceptively that the premises should have been reconsidered in the first place.

Sticking for the moment to the notion of rationality, it is clear that this notion needs to be reconceptualized in order to move out of the unpleasant choice between absolutism and relativism. We should avoid attributing special qualities to rationality, that are required to hold in each and every historical context. But this should not lead to the complete exclusion of rationality as a determining factor, as some have argued.[82] Possible fruitful areas to investigate in this respect are the work of the later Kuhn and some interpretations of the evolutionary epistemology programme. However, one of the most interesting recent developments is the research on rationality in terms of coping with uncertainty that has been conducted in cognitive psychology.[83] Gigerenzer, for example, defends the claim that rationality should be seen as a kind of heuristics. This means that forms of reasoning are apt only for specific type of problems in specific types of situations. There is not one overriding concept of rationality, instead a plural approach is desired. Yet this does not render historical analysis fully dependent on the past as types of situations can occur in more than one context. This angle can be highly relevant for the study of errors in science, especially since a connection between uncertainty, types of situations and rationality is brought about in it.[84]

These lines for future research offer a way out of the unworkable dichotomy between errors as obstacles and errors as failures. In any case the goods points of the two main approaches should be retained and this appears to be possible only when the suggested ideas for future research are followed. From the 'errors as failures' approach we must retain the demand to stay close to local historical contexts. Contextual investigations are necessary in order to account for the diverse conceptual and physical circumstances in which errors originate and offer explanations for the sources of error and the ways in which they were dealt with. Moreover the 'errors as failures' approach has shown that what counts as an error depends on its relation to some frame of reference. The drawback was that such frames of reference could only be captured in highly specific socio-cultural terms. We must also drop the strict division between internal and external factors in explaining scientific products. From the 'errors as obstacles' approach we must retain the traditional evaluative stance with respect to error and truth and take seriously the need to make use of diachronic historical perspectives. Diachronic historiography can only be predicated on assumptions of continuity. The work within 'New Experimentalism' on error, but also the practical turn in historiography can be very useful here because they both assume continuity in experimental practices.

Historiography of science must be seen as an exceptional field of historical study. Part of the subject matter the historian of science addresses is qualitatively different from other fields of historical inquiry. Rules in science are expected to hold everywhere and no exceptions to them are allowed. Rules of social conduct, on the other hand, do allow for exceptions, deviations and so forth.[85] This means that errors made in science have a significantly different consequence than errors made in warfare, political decision-making and so on. Hegel famously declared that the only thing that we learn from history is that we do not learn from history. The history of science might well be the single exception to this rule, if of course a satisfactory theory of error and hence a satisfactory theory of learning can be provided.

2 EXPERIMENTAL KNOWLEDGE IN THE FACE OF THEORETICAL ERROR

Kent W. Staley

Introduction

Much of the advancement of scientific knowledge is due to or consists of advances in experimental knowledge. Such experimental progress, moreover, owes a great deal to advances in our understanding of how to systematize, quantify and eliminate error. It is therefore no surprise that the turn by philosophers of science toward error as a subject of interest in its own right is strongly associated with their turn toward experiment, beginning with the 'New Experimentalists' in the 1980s. Much of the work that characterized the New Experimentalism in philosophy of science emphasized the idea that, in Hacking's words, 'experimentation has a life of its own', a slogan that could be interpreted differently depending on whether one wished to emphasize the continuity of experimental methods across radical changes in theory,[1] the possible independence of experimental evidence from theoretical assumptions,[2] or the growth of knowledge at the experimental rather than the theoretical level as the driver of scientific advancement.[3]

It was perhaps inevitable that there would be some backlash against this emphasis on the independence of experiment from theory. Even authors who contributed to the turn toward experiment began to point out the important role played by theoretical assumptions in arriving at experimental results.[4] Recognizing this role, however, makes it incumbent upon philosophers of science to gain clarity about the effects it might have upon the possibility of cumulative knowledge at the experimental level (i.e. both knowledge gained by inferences made from experimental data and knowledge about how to reliably produce, control and analyse experimental data).

The present chapter attempts to make progress toward a greater understanding of the problem of theoretical error in the context of drawing conclusions from experimental data. More precisely, my discussion will focus on the problem of theoretical assumptions that threaten to lead investigators to either misinterpret their data or to incorrectly evaluate the support the data provide for their conclusions. Such erroneous theoretical assumptions demand philosophical attention because unlike, say, possible error due to variability in a sample, we presently have no satisfactory quantitative statistical framework for taking such errors into account. I will consider a number of different ways in which errors in theoretical assumptions can manifest themselves at the empirical level. Theoretical assumptions pose dangers when drawing inferences from experimental data for several reasons. Here I will discuss examples highlighting problems that fall into two broad categories: (1) *problems of scope*: because of their great generality theoretical assumptions are difficult to test exhaustively and sometimes must be applied in domains far from those in which they have previously been tested; (2) *problems of stability*: the rare-but-recurring historical phenomenon of unforeseen and dramatic *change* in general explanatory theories arguably constitutes some reason to worry about similar upheavals overturning theoretical assumptions that presently seem sound; moreover, conceptual innovation sometimes enables erroneous *implicit* assumptions to be brought to light.

In the face of these dangers, I will defend a guardedly optimistic position: experimenters can avail themselves of methods that allow them to secure their experimental conclusions against the possibilities of erroneous theoretical assumptions, provided they confine themselves to certain kinds of conclusions, i.e. those that I shall characterize as resting upon secure premises. Through two episodes in the history of gravity physics, I show how physicists have developed strategies for securing their inferences regarding gravity against possibilities of error. I argue that these strategies should be construed as directed at *epistemically possible error scenarios*, i.e. ways that the world might be in an epistemically modal sense. Such strategies are central to the *secure evidence framework*.

The plan of the chapter is as follows: in the next section I lay out the secure evidence framework for discussing the possibilities of error in drawing conclusions from experimental data. Within that framework, three types of strategies can be distinguished for securing evidence: strengthening, weakening and robustness. The next two sections survey episodes from the history of gravity physics that exemplify these strategies and show how physicists have used them to confront the problem of relying on possibly erroneous theoretical assumptions. The section on pp. 44–7 discusses the strategies used by Newton to reason securely about gravity from the data available to him. In the section that follows this (pp. 48–54), I discuss the successors to Newton who are using a parametric framework to explore systematically the possibilities of error in general relativity. I clarify the theoretical assumptions behind that framework, and show how physicists are able to deploy still further advances in the ability to relate data to fundamental physics to secure those theoretical assumptions. I also show that robustness plays a role in understanding how the conclusions they draw are compatible with the possibilities of error that remain in those theoretical assumptions. I summarize my conclusions in the last section (pp. 54–5).

Securing Evidence

The issues that are of concern here relate to at least two central problems for experimental science. (1) Given a body of data, what propositions may justifiably be inferred from those data? Alternatively, for what hypotheses do those data serve as good evidence? (2) Given a number of hypotheses of interest, which constitute potential answers to some question, how might one go about generating data that could serve as good evidence for or against one or more of those hypotheses?

For either question, the experimenter must consider what she may safely assume to be true. Even for purely statistical inferences from data, one must rely on assumptions that function as additional premises. Such assumptions might consist of invoking a statistical model of the data-generating process (as in the case of, for example, an inference using Neyman–Pearson statistics), or of a prior probability distribution across the hypotheses of interest (as in a Bayesian inference). Such statistical assumptions might themselves be founded upon material assumptions, as varied as the experimental techniques themselves, about the conditions under which data are gathered: are the reagents sufficiently pure? Is the shielding against cosmic rays sufficient? Are the responses of interview subjects influenced by the order in which the questions are asked? Are there dietary differences between the control and treatment groups that contribute to any apparent difference in outcomes?

We must note at the outset the distinction between the exact truth of the explicit statement of an assumption and the *adequacy* of such a precisely stated assumption. That, for example, survey responses are independent of question order is a statement that can, in a given setting, be rendered with considerable precision. In practice, experimenters are often able to rely on such statements without them being, in the sense of that precise statement, exactly true. If survey response has a sufficiently weak dependence on question order, then perhaps one can simply ignore the dependence and proceed *as if* it did not exist at all.[5] Even so, truth remains relevant. What is being assumed to be true is not the precise statement 'survey response is independent of question order', but a less precise statement such as 'survey response may safely be taken, for present purposes, to be independent of question order'.

We can thus articulate a little more carefully the problem with which we are concerned. The problem of 'theoretical error' is the problem that, in drawing inferences from data, investigators might draw upon theoretical assumptions that are *inadequate* for the inference at hand, in the sense that those assumptions are likely to lead investigators to misinterpret the data or to mistake the extent to which the data support their inferences.

Before moving forward, it will be helpful to distinguish between two phases or modes of reasoning with regard to experimental data. In what we might call the *use mode*, scientists use assumptions, including theoretical assumptions, to arrive at substantive conclusions from experimental data. In the *critical mode*, scientists turn their attention to those assumptions themselves, subjecting them to testing and criticism of various sorts. In the case of model-based statistical inference (for example, Neyman–Pearson inference), this distinction corresponds to that between the primary inference, which is based on an assumed model, and 'model criticism', which is concerned with the statistical adequacy of that model.[6] At least implicitly, the use of a set of assumptions in arriving at substantive conclusions commits one to the claim that one knows enough for inferences based on those assumptions to be reliable. The critical mode is crucial for warranting such a commitment.

I wish to characterize the epistemic function of the critical mode of reasoning in experimental contexts in terms of the *securing of evidence*, a term that I will next undertake to explain.

The investigator who turns a critical eye toward the assumptions employed in her inferences from data must be concerned with the state of her own knowledge and the constraints it sets upon the ways in which her inferences might go wrong. We can think of her as contemplating the ways the world might be that would, were they actual, result in the falsehood of her claims about what the data support.

Here, the relevant modality of the word 'might' must be an epistemic modality. Merely logical possibilities of error can be safely ignored if one knows that they are not actual. The same can be said for counterfactual possibilities, while what philosophers like to call 'metaphysical possibility' is simply too remote from the concerns of empirical science to be relevant to this question. There is an emerging literature and an ongoing disagreement about the semantics of epistemic modal propositions.[7] Fortunately, that debate can safely be ignored here. The relevant feature of epistemic modality for our purposes is that what is epistemically possible in epistemic situation K is limited by the propositions that are known in K. More precisely, gaining knowledge renders scenarios that were possible, impossible.[8] Conversely, knowing less means confronting a greater range of possibilities.[9]

In what follows I will apply the concept of epistemic possibility to *error scenarios*, and here I explain that term. If C is an evidence claim (i.e. a statement expressing a proposition of the form 'data x from test T are evidence for hypothesis H'), and S is an epistemic agent contemplating asserting C whose epistemic situation (her situation regarding what she knows, believes, is able to infer from what she knows and believes, and her access to information) is K, then a way the world might be, relative to K, such that C is false, we will call an *error scenario* for C relative to K.

It is common in scientific discourse to find investigators, when presenting and justifying their inferences from data, addressing directly the question of what error scenarios might and might not be possible, given their epistemic situation (typically this is the epistemic situation of a collective agent, such as a research team). I propose that we can conceptualize the justificatory function of such practices with the help of the notion of security.

Suppose that, relative to a certain epistemic situation K, there is a set of scenarios that are epistemically possible, and call that set Ω_0. If proposition P is true in every scenario in the range Ω_0, then P is *fully secure* relative to K. If P is true across some more limited portion Ω_1 of Ω_0 (i.e. $\Omega_1 \subseteq \Omega_0$), then P is secure throughout Ω_1.

To put this notion more intuitively, then, a proposition is secure for an epistemic agent just insofar as that proposition remains true, whatever might be the case for that agent. Thus defined, security applies to any proposition, but the application of interest here is to evidence claims and inferences. Specifically, an *inference* from fact e to hypothesis h is secure relative to K insofar as the proposition 'e is good evidence for h' is secure relative to K.

It is important to stress that the methodological benefit of the security concept derives not from full security but rather from the ways in which various practices serve to *increase relative security*. I do not suppose that inquirers are ever called upon to determine the degree of security of any of their inferences. The methodologically significant concept turns out to be not security *per se*, but the *securing* of inferences, i.e. those practices that increase the relative security of an evidence claim.

Such practices can be classified as falling into two broad types of strategy: *weakening* and *strengthening*. In weakening, the conclusion of an evidential inference is logically weakened in such a way as to remain true across a broader range of epistemically possible scenarios than the original conclusion. Strengthening strategies operate by adding to knowledge, reducing the overall space of epistemically possible scenarios so as to eliminate some in which the conclusion of the evidential inference would be false.

A third security related justificatory strategy, which we might label *robustness*, aims at the analysis and clarification of the security of an evidence claim, by showing, for some class of error scenarios, that an evidence claim put forth remains true in those scenarios. This strategy is often used in combination with either weakening or strengthening strategies. A significant literature has grown up around the issue of robustness, and the term has been given a variety of interpretations.[10] My use of it here is restricted: the robustness strategy with regard to security is a strategy whereby one clarifies the justification of an evidence claim by showing that the claim remains true across some range of error scenarios. I have previously argued that robustness of evidence can be epistemically valuable by contributing to the security of an evidence claim.[11]

In what follows I will discuss experimental undertakings in the face of uncertainty about theoretical uncertainty that illustrate the three strategies just mentioned.

Newton's Arguments for the Law of Universal Gravity: Strengthening and Weakening Strategies

In Book 3 of *Principia*, Newton lays out a step-by-step argument for his universal law of gravitation. Beginning with 'phenomena' that describe the ways in which the planets and their respective satellites satisfy Kepler's laws, Newton argues for a series of propositions, leading ultimately to the claim that 'gravity exists in all bodies universally and is proportional to the quantity of matter in each' (Proposition 8), and that this gravitational force is a force of attraction that is inversely proportional to the square of the distances between the centers of the bodies in question.[12]

Here I do not propose to recapitulate Newton's primary positive argument for the universal law of gravity, which can be understood as having been advanced in the 'use mode' referred to above.[13] Rather, I wish to consider Newton's means, in the 'critical mode', of securing the conclusion of that argument against possible error scenarios. Some of his responses to possible error scenarios draw upon aspects of his primary positive argument, but others require him to go beyond that argument.

First, could the universal law of gravitation be subject to a large error in the direction of the attractive gravitational force? This error Newton rules out on the basis of the fact that the rate at which areas are swept out by orbital radii can serve (Kepler's 'harmonic law') as a *measurement* of the direction of the attractive force.[14] Second, could the law be subject to a large error in its specification of the distance-dependence of attraction? Again, Newton can use a theory mediated measurement as the basis for his rejection of this possibility: according to Corollary 1 of Proposition 45 in Book 1, the power of the distance dependence is measured by the precession of the moon's orbit around the Earth.[15]

These points are crucial to Newton's primary argument for the law of gravity. By appealing to such methods, Newton is able to argue that these phenomena do constitute good evidence for the law of gravity. What I would like to consider a little more closely is how Newton, having made that argument, secures his inference against error scenarios that threaten the underlying theoretical premises of the argument.

The first such challenge was raised by Cotes. Cotes challenged Newton's application of the third law of motion (that 'the actions of two bodies upon each other are always equal and always opposite in direction') to gravitation.[16] Granting that one may justifiably apply the third law to contact forces, can one apply it to forces of attraction relating two bodies separated in space? If not, then this would undermine Newton's inference that makes gravitation universal: perhaps not all bodies exert gravitational forces upon one another. Cotes expressed this concern with the help of an imaginary scenario:

Suppose two Globes *A* & *B* placed at a distance from each other upon a Table, & that whilst *A* remains at rest *B* is moved towards it by an invisible Hand. A bystander who observes this motion but not the cause of it, will say that *B* does certainly tend to the centre of *A*, & thereupon he may call the force of the invisible Hand the Centripetal force of *B*, or the Attraction of *A* since ye effect appears the same as if it did truly proceed from a proper & real Attraction of *A*. But then I think he cannot by virtue of the Axiom [Attractio omnis mutua est] conclude contrary to his Sense & Observation, that the Globe *A* does also move towards the Globe *B* & will meet it at the common centre of Gravity of both Bodies. [17]

Cotes's proposes an error scenario that threatens Newton's inference by calling into question a theoretical assumption about the applicability of the third law of motion to forces of attraction between bodies that are not in contact. The error scenario in question holds that what appear to be motions produced by mutually acting centripetal forces are really produced by mechanical pushes on the orbiting body. Since the appearance of a force exerted upon one body by another is, under this scenario, an illusion, the third law does not apply.

Were Newton to simply assume hypothetically that the third law applies in this case, as alleged by Stein,[18] his inference would fail to be secure against this error scenario. As shown by William Harper, however, Newton does have resources for answering Cotes's challenge. By appealing to these resources, Newton is able to strengthen his epistemic situation, ruling out the kind of error scenario invoked by Cotes.[19] I will not attempt here to rehearse Harper's account of Newton's response to Cotes's challenge in its entirety, nor to explain the arguments in detail. Nonetheless, what follows will, I hope, suffice to establish that Newton is able to use a strengthening strategy to secure his inference against this particular error scenario.

First, Newton is able to appeal to other kinds of attractive forces between spatially separated bodies that evidently *do* exert equal and opposite forces upon one another: namely, magnetic attractions between a lodestone and a sample of iron, as Newton found in his own experiments.[20] Even if these motions are the result of some kind of ether, Harper argues, as the mechanical philosophy would require, the mutual endeavour of the two bodies to approach one another would still be subject to the third law. Cotes's invisible hand does not qualify as such a mutual endeavour.

Newton's second argument appeals to the gravitational equilibrium of the parts of the Earth itself. Were the attractive forces exerted by, say, one hemisphere upon the other, not equal and oppositely directed, then there would result a net acceleration of the planet as a whole.[21]

Finally, Harper explains how applying the third law of motion to the Sun and Jupiter leads to convergent results when combining independently evaluated acceleration fields. In a nutshell, the argument goes as follows: Newton can use estimates of the distances between the Sun and the planets to estimate the centripetal acceleration field directed toward the Sun, yielding an estimate of Jupiter's acceleration towards the Sun. Likewise Newton can use James Pound's data for Jupiter's four moons to estimate the acceleration field centred on Jupiter. Using the third law to extrapolate this field to the distance separating the Sun from Jupiter yields an estimate of the Sun's acceleration towards Jupiter. Using the ratios of these two quantities, one can then define a common centre of rotation about which the Sun and Jupiter will orbit with a common period. This yields measures of the weight of each of the two bodies towards the other, weights that remain oppositely directed to each other as the two bodies orbit their common centre. As Harper writes,

> They, therefore, fulfill one major criterion distinguishing what Newton counts as attraction from Cotes' invisible hand pushing one body toward another. To have these oppositely directed weights count as a single endeavor of these bodies to approach one another requires, in addition, that they be equal so that they satisfy Law 3.[22]

Newton thus shows how the assumption that the third law does apply to gravitational accelerations leads to convergent estimates of acceleration fields based on distinct bodies of data, adding, according to his own methodology, to the evidence for the premises of his main argument.

Next I would like to show how Newton uses a weakening strategy to respond to a different kind of challenge. In a weakening strategy, the investigator weakens the conclusion of an inference, or opts for the weaker of two possible conclusions to an inference, so that an error scenario that would threaten to undermine the inference to a stronger conclusion is no longer threatening. The scenario may remain possible, but even if it is actual the inference remains probative.

Consider the following scenario: maybe it is changes in motion of invisible particles in a vortex, such as that postulated by Huygens, that pushes the planets into orbital motion, rather than mutual gravitation of the Sun and planets.

The first thing to be said here is that, as Harper explains, Huygens's theory itself predicted that the planets would follow exact Keplerian orbits. As such the theory was only successful insofar as it predicted orbits consistent with the data available at the time. But what Harper calls Newton's 'richer' ideal of empirical success requires more than this. Newton's application of the third law of motion, for example, yielded converging measurements of relative inertial masses in the solar system, something that Huygens's theory was unable to accomplish. Later work demonstrating perturbations to Keplerian orbits for solar system bodies, beginning in the 1740s but becoming truly compelling only in 1785 with Laplace's 'Théorie de Jupiter et de Saturne', subsequently eliminated Huygens's theory decisively (though it had ceased to attract adherents well before this).[23]

But Newton's weakening strategy is a response to a less well-defined challenge. This strategy is essentially captured in his famous 'hypotheses non fingo'.
The challenge takes the form of the following potential error scenario (or more
precisely, a class of potential error scenarios specified only by a feature that they
all share): Perhaps somehow etherial particles act in such a way as to produce the
mutual attractions that Newton describes using the law of gravity.

Here Newton's weakening strategy is his decision to opt for a conclusion
from the data available to him that is weak enough to be compatible with this
possibility. I refer to this as a *potential* error scenario because it would be an error
scenario were Newton to draw the stronger conclusion from his data that bodies
endeavour to approach one another with forces that are attributable to gravitational attraction *and nothing else*. That he does not intend for the universal law
of gravity to be given such a strong interpretation is clearly indicated by his oft-
quoted statement that 'I have not as yet been able to deduce from phenomena
the reason for these properties of gravity, and I do not feign hypotheses'.[24] In
other words, his conclusion should be understood as 'Gravity really exists and
acts according to the laws that we have set forth and is sufficient to explain all the
motions of the heavenly bodies and of our sea', rather than as, 'Gravity consists of
the mutual attractions of bodies acting upon one another at a distance in a way
that does not involve locally acting causes (such as etherial particles)'.

To be sure, Newton's refusal to draw this stronger conclusion has an independent motivation insofar as he believes it would commit him to regarding
gravitational attractions as instances of action at a distance, which, as expressed
in a letter to the clergyman Richard Bentley, he regarded as an 'absurdity'.[25] Even
had he not held this commitment, however, it would remain true that such a
stronger conclusion would have been less warranted, precisely because he would
not have been able to rule out the potential error scenario in question.

Summary

Newton faced a significant problem of scope with regard to his application of the
third law of motion to gravitational forces, because these forces acted between
bodies that were not in contact and because it was difficult to test directly
the hypothesis that the third law applied to gravitational attractions between
celestial bodies. Newton was able to secure his assumption by appealing to the
gravitational equilibrium of the parts of the Earth and to the agreeing estimates
of centripetal acceleration fields that resulted from thus applying the third law
of motion. He could also appeal to the third law's applicability to similar ter-
restrial phenomena involving magnetism. Finally, he drew a conclusion that was
sufficiently weak to remain compatible with possible mechanical explanations of
the cause of gravity.

Parametric Frameworks for Testing Fundamental Physics: Strengthening and Robustness

The fate of Newton's conception of gravity as it was supplanted by Einstein's general relativity exemplifies the problem of stability. That gravity should be thought of in terms of the curvature of space-time rather than forces of mutual attraction is clearly an error scenario against which Newton's inferences were not secured, even though, as Harper argues, Newton's own methodology provided the engine of his own theory's demise.[26] In this section I will discuss the ways in which recent gravity researchers have secured their inferences regarding modern theories of gravity against error using a more systematic approach that constitutes a striking methodological advance. I will also argue that the implementation of the methods I describe involves ruling out possible errors in the assumptions of that method, and that the appropriate conception of the possibilities thus invoked is epistemic.

The parametrized post-Newtonian (PPN) formalism was developed to enable the comparison of metric theories of gravity with each other and with the outcomes of experiment, at least insofar as those theories are considered in the slow-motion, weak-field limit. Metric theories of gravity can be characterized by three postulates:

1. space-time is endowed with a metric **g**,
2. the world lines of test bodies are geodesics of that metric, and
3. in local freely falling frames (Lorentz frames) the nongravitational laws of physics are those of special relativity.[27]

The PPN approach facilitates comparison of such theories using a common framework for writing out the metric **g** as an expansion, such that different theories are manifested by their differing values for the constants used in the expansion. As Clifford Will writes,

> The only way that one metric theory differs from another is in the numerical values of the coefficients that appear in front of the metric potentials. The [PPN] formalism inserts parameters in place of these coefficients, parameters whose values depend on the theory under study.[28]

Using her error-statistical framework, Deborah Mayo has emphasized the positive role played by the PPN framework in facilitating, not only the comparison of existing theories, but also the construction of new alternatives as a means of probing the various ways in which general relativity (GR) could be in error. In addition, she argues that the resulting proliferation of alternatives to GR was not a manifestation of a theory in 'crisis', but rather of an exciting new ability to probe gravitational phenomena and prevent the premature acceptance of GR. A key to the strength of this approach is the way in which the PPN formalism allows for the combination of the results of piecemeal hypothesis tests, not only to show that some possibilities have been eliminated, but to indicate in a positive sense the extent to which and ways in which gravitation is a phenomenon that GR (or theories similar to GR) gets right (see Table 2.1).[29]

Table 2.1: The PPN parameters

Parameter	What it measures relative to GR	Value in GR
γ	How much space-curvature produced by unit rest mass?	1
β	How much 'nonlinearity' in the superposition law for gravity?	1
ξ	Preferred-location effects?	0
α_1	Preferred-frame effects?	0
α_2		0
α_3		0
ζ_1	Violation of conservation of total momentum?	0
ζ_2		0
ζ_3		0
ζ_4		0

Source: Adapted from Will, 'The Confrontation Between General Relativity and Experiment'.

To take just one example of this use of the PPN framework to search for possible departures from the predictions of GR and thus set limits on the extent to which GR could be mistaken, consider the use of very long baseline interferometry (VLBI) to measure the PPN parameter λ, which measures the curvature of space. The relationship between λ and the predicted angle of deflection θ for an electromagnetic ray from a distant source due to the Sun is given by

$$\theta \cong \frac{(1+\lambda)GM_\odot}{c^2 b}(1 + \cos\phi), \tag{1}$$

where G is the gravitational constant, c is the speed of light in a vacuum, $M_\odot$ is the mass of the Sun, b is the distance of closest approach from the centre of the Sun to the ray's path, and ϕ is the angle between the source of the ray and the Sun as viewed from Earth.

General relativity, in which the only dynamical field is the metric $\mathbf{g}$, assigns a value $\lambda = 1$, but some alternative theories of gravities include additional parameters that allow λ to take other values. In scalar-tensor theories the matter fields couple not only to the metric but also to an additional gravitational scalar field. In Brans–Dicke theory, for example, the effects of this additional scalar field show up in the form of an additional parameter ω_{BD}, such that $\lambda = (1 + \omega_{BD})/(2 + \omega_{BD})$. The greater the effects of the scalar field, the smaller the value of ω_{BD}, and the greater the departure of λ from unity.

Interferometry, which has been the basis of crucial experimental insights going back to the Michelson–Morley experiment, looks for differences in times of arrival for light signals originating from a single source that have travelled different paths. VLBI employs telescopes at different locations around the Earth equipped with local atomic clocks as timing devices to allow for the combination of data from widely separated detectors. The idea behind using VLBI to measure λ, first suggested by Irwin Shapiro,[30] is to apply such interferometric techniques to light from celestial radio wave sources.

As one example of such an experiment, consider the results reported by S. S. Shapiro, Davis, Lebach and Gregory.[31] They base an estimate of λ on twenty years' worth of VLBI data collected from 1979 to 1999, with signals from 541 sources producing data at 87 VLBI sites. This yielded 1.7×10^6 measurements. Bypassing the somewhat intricate error analysis attempting to take into account possible errors in assumptions regarding matters such as atmospheric propagation delay and atmospheric refractivity gradients, I will simply note that they report an estimate of $\lambda = 0.9998_3 \pm 0.0004_5$, which is 'within one standard deviation of the value predicted by GR'.[32] (A value of $\lambda = 0.9994$ corresponds to $\omega_{BD} = 1.665 \times 10^3$.)

As Will notes, for large values of ω, scalar-tensor theories make predictions for the current epoch that agree with those of GR 'for all gravitational situations – post-Newtonian limit, neutron stars, black holes, gravitational radiation, cosmology' to within the order of ω^{-1}.[33] What Shapiro et al. thus allow us to conclude is that for all these phenomena, any effects of a scalar gravitational field, should there be one, will be such that we can rely on GR to give us the correct prediction to at least one part in a thousand.

That is what we may conclude, in the use mode, supposing that we may safely accept the assumptions of the PPN framework itself, from which the parameter measured in the experiment performed by Shapiro et al. derives its meaning. Recall that the PPN framework encompasses only *metric* theories of gravity. Such theories, which treat gravity as a manifestation of curved space-time, satisfy the Einstein equivalence principle (EEP). EEP is in turn equivalent to the conjunction of three apparently distinct principles – local position invariance (LPI), local Lorentz invariance (LLI) and the weak equivalence principle (WEP).[34] Thus, experimental conclusions placing limits on the values of the PPN parameters rest upon a substantial theoretical assumption: the EEP. In a shift from the use mode to the critical mode, Will writes, 'The structure of the PPN formalism is an assumption about the nature of gravity that, while seemingly compelling, could be incorrect'.[35]

Will's comment involves two assertions that both demand attention in order to see how experimental knowledge can be secure in the face of possible theoretical error. First, he claims that EEP is 'seemingly compelling'. Second, he admits that it 'could be incorrect'.

Taking these claims in order, it first must be clarified that Will's use of the term 'seemingly' is not meant to suggest that the support for the EEP is merely illusory. Indeed, he elsewhere writes that there is sufficient evidence that 'gravitation ... must be described by a "metric theory" of gravity'[36] and that this evidence 'supports the conclusion that the only theories of gravity that have a hope of being viable are metric theories, or possibly theories that are metric apart from very weak or short-range non-metric couplings (as in string theory)'.[37] (I comment below on the latter qualification.) It is evident that Will maintains that we may, at least for the purposes of using the PPN parameters, safely assume the EEP. The basis for this claim is that the EEP itself has been subjected to some very high precision tests that can be utilized in the context of other parametric frameworks that facilitate the testing of WEP, LLI and LPI.

For example, in 1973, Lightman and Lee developed the *THεμ* formalism, which functions analogously to the PPN framework for tests of GR.[38] The class of theories that can be described within the *THεμ* formalism includes all metric theories. It also includes many, but not all, non-metric theories.[39] The ability to put non-metric theories into a common framework such that limitations can be put on EEP violations in a systematic way provides a powerful extension of the programme of testing within PPN.

This formalism has proven to be adaptable to the pursuit of tests of null hypotheses for each of the components of EEP. By taking various combinations of the four *THεμ* parameters, one can define three 'non-metric parameters', Γ_0, Λ_0, and Υ_0, such that if EEP is satisfied then $\Gamma_0 = \Lambda_0 = \Upsilon_0 = 0$ everywhere. Tests of the components of EEP can then be investigated in terms of null tests for these parameters. A non-zero value for $_0$ is a sign, for example, of a failure of LLI. Will describes how the results of the Hughes–Drever experiment ('the most precise null experiment ever performed'),[40] can be analysed so as to yield an upper bound of $\Upsilon_0 < 10^{-13}$ and concludes that 'to within at least a part in 10^{13}, Local Lorentz Invariance is valid'.[41] Eötvös experiments have tested WEP and yielded limits on 'non-metric parameters' of $|\Gamma_0| < 2 \times 10^{-10}$ and $|\Lambda_0| < 3 \times 10^{-6}$.

More recent tests of Lorentz invariance have employed another formalism, the standard model extension (SME).[42] Because violations of Lorentz invariance are expected in a number of proposals for physics beyond the standard model (SM), experimental tests that might reveal such violations without having to assume any particular non-SM dynamics offer the possibility of insights into such daunting theoretical problems as quantum gravity. The SME uses an effective field theory approach to generalize the SM and GR so as to allow for the quantitative description of violations of both LLI and CPT invariance, adding new terms to the SM that allow for CPT and Lorentz violation in different sectors. Experimental tests can then be used to set limits on the coefficients for these terms. For example a clock-comparison experiment like the Hughes–Drever experiment looks for variation in the frequency of a clock (typically an atomic transition frequency) as its orientation changes. A descendant of the Hughes–Drever experiment, using a ^{3}He/^{129}Xe maser system, has yielded a constraint on a combination of the Lorentz-violating coefficients in the neutron sector of the SME (coefficients characterizing a possible Lorentz-violating coupling of neutrons to a possible background tensor field traceable to spontaneous symmetry breaking in a fundamental theory that need not be specified) at the level of 10^{-32}.[43]

It is this test and myriad other high-precision tests of the components of EEP[44] that stand behind Will's assertion of the 'seemingly compelling' evidence for EEP. We must now turn to his comment that, nonetheless EEP 'could be incorrect'. An alternative formulation of this statement is: 'It is possible that the EEP is false'. But in what sense of 'possible'?

We can quickly dispense with an easy question: is it logically possible that the EEP is false? Surely, it is. (More carefully, we haven't any reason to think that the denial of EEP in some hidden way harbours a logical contradiction.) But that is not relevant to whether we may safely assume EEP to be true. We can also quickly dispense with a difficult question: is it physically possible that EEP is false? Assuming the standard view that what is physically possible is whatever is not ruled out by the laws of physics, the answer is that we do not know. It is precisely to find out whether violations of EEP are permitted by the laws of physics that we wish to test the EEP.

I would suggest that the important sense in which we need to consider the possibility of the failure of EEP is an epistemic one: given what we do know, does it remain possible that the EEP is false? Not, presumably, if we know that EEP is true. This would constitute an application of what is sometimes called 'Moore's Principle', which states that if a speaker truthfully asserts that she knows that a certain proposition P is true, then she is not in a position to assert that P might be false.[45] I do not propose here to defend the truth of Moore's Principle, but only wish to claim that, provided that the modality implicated is epistemic, an assertion by S that P might be false is seriously in tension with S's truthful assertion that she knows P.

However, there is a difference between knowing that EEP has been well-tested and knowing that it is true full-stop. There are, as it turns out, two distinct bases for the statement that 'EEP might be false'. One basis is that of the determined sceptic, and falls prey to all of the difficulties that the sceptic faces. The second is more scientifically relevant, but poses no threat to the possibility of gaining experimental knowledge about gravity by means of tests of hypotheses of the PPN parameters.

First, there is the sceptic's gambit: given any finite body of data that appears to lend support to a theory, it is always possible to come up with some other theory that can be made compatible with those data. To this one appropriate response is to say 'Yes, but so what?' From Newton's fourth rule of reasoning[46] to Peirce's injunction that one must not 'block the road of inquiry' to Deborah Mayo's critique of 'gellerized' alternative hypotheses,[47] a long tradition in philosophy of science has equipped us to provide a methodological critique of the strategy of avoiding experimental conclusions by invoking the mere possibility of alternative hypotheses. Here, Mayo's criticism will serve us nicely: the strategy is unreliable.[48] More specifically, a person who is determined to avoid conclusions drawn from data because of the possibility of alternative hypotheses that can be constructed after the fact to fit those data will always fail to accept hypotheses that are true.

But, inspired perhaps by Moore's Principle cited above, one might object that admitting an alternative to EEP, compatible with all of the data generated in tests of EEP, to be possible amounts to admitting that we do not know EEP is true, and thus we cannot rely on EEP in the interpretation of tests conducted within the PPN framework. Clifford Will himself has written that 'Nothing can be categorical, of course, because given any finite experimental error, one could always in principle conceive of a non-metric theory that satisfies all tests of EEP' (personal correspondence). Here there is a danger of landing in a philosophical muddle. We have conflicting accounts of knowledge and of epistemic possibility. Are we doomed to similar conflict over the possibility of knowledge about gravity?

The first reason I think that we are not thus doomed is that we have good reason to dismiss most such alternatives, which would have to be so extremely gerrymandered as to be utterly implausible.[49] These alternatives fall into the category of 'conspiracy theories' – effectively, they suppose that Nature has gone to great lengths to make things look as if EEP is satisfied everywhere that we do look, and hides the exceptions from us very cleverly.

The second reason we are not doomed is closely tied to the second basis for the admission that, after all, EEP 'could be incorrect'. We in fact do not know, and, for the purposes of interpreting PPN experiments, do not need to know that EEP is *exactly* correct. Many physicists expect EEP in some way or other to fail to hold in the regime of quantum gravity. Indeed, the very meaning of what a metric theory is becomes unclear in these contexts. Clifford Will notes that string-inspired theories

> are 'metric' in a deep sense, because they are built upon a metric foundation; however they can have additional fields (dilatons, moduli) that couple to matter in a non-universal way, and can lead to violations of EEP. One line of reasoning would treat these fields as gravitational fields, in the same class as the metric, and thus would call these theories non-metric. But another viewpoint would treat these fields as additional (admittedly exotic) MATTER fields, no different, really, from the electromagnetic field, which obviously couples to matter in a non-universal way. (personal correspondence)

PPN allows us to restrict our attention to settings in which these kinds of concerns can be set aside.

This final point can be appreciated as an application of a robustness strategy. There are indeed error scenarios in which the EEP fails to hold, but because the PPN is meant to capture the description of gravitational phenomena in the slow-motion, weak-field domain in which such violations will not be manifest, the conclusions drawn via PPN would remain the same across such error scenarios.

Summary

The PPN framework succeeds in part by restricting its attention to only metric theories of gravity, which amounts to assuming the EEP, which might not be true. Securing the PPN results that allow us to draw conclusions about gravity draws upon both strengthening and robustness strategies. The strengthening strategy uses tests of the EEP commitments to eliminate possible scenarios that would undermine the conclusions drawn in the PPN framework. There are, however, possible scenarios, and serious ones, in which the EEP fails to hold exactly, or in which it is unclear what the EEP requires. But the conclusions drawn about the values of the PPN parameters remain valid in all of these scenarios, which do not make a difference to the slow-motion, weak-field approximation on which the PPN is based. This latter point invokes robustness as a means of safeguarding experimental knowledge about gravity against possibly erroneous theoretical assumptions.

Conclusion

These developments in the experimental study of gravity exemplify the importance of theoretical assumptions in drawing conclusions from experimental data. They also show that efforts to secure those assumptions so that investigators have the epistemic resources to reason from safely held assumptions are equally a part of the legacy of experimental physics, as they presumably are of other experimental sciences.

If the challenges of reasoning from experimental data in the face of uncertain theoretical assumptions have been more or less constant, the resources developed for confronting those challenges have hardly been stagnant. The proliferation of parametric frameworks for bringing experimental data into contact with fundamental physical theories constitutes a recent example of *progress* at the interstices between experimental technique and theoretical elaboration. That such progress in the methodology of bringing data to bear on theory has a life of *its* own, or at least a history to be explored, is a promising prospect for historians and philosophers of the experimental sciences.

I have also argued that these efforts to secure inferences about gravity against possibilities of error should be construed as directed at error scenarios that are possible in the epistemic sense, as opposed to other modalities that have preoccupied philosophers. A recent surge of philosophical interest in epistemic modality has occurred independently of considerations in the philosophy of science and without the participation of philosophers of science.[50] The present discussion serves, however, to demonstrate the relevance of epistemic modality to understanding how scientific investigators warrant inferences from data.

3 LEARNING FROM ERROR: HOW EXPERIMENT GETS A LIFE (OF ITS OWN)

Deborah G. Mayo

'Experiment Lives a Life of its Own!'

This familiar slogan is often thought to capture the epistemic importance and power attributed to experiment. But decades after Hacking[1] and others popularized it in philosophical circles, we are still in need of building a full-bodied philosophy of experiment.

- How should we understand this slogan?
- Where does the separate life of experiment take place?
- How does it manage to get a life of its own?
- Why should it want its own life?

Beginning with the first question, we may consider three interrelated glosses on the 'own life' slogan.

Experimental Aims: Apart from Theory Appraisal

The first sense in which experiment may be said to have its own life concerns experimental aims: to find things out quite apart from testing or appraising any theory. The goals are the local ones of obtaining, modelling and learning from experimental data: checking instruments, ruling out extraneous factors, getting accuracy estimates, distinguishing real effect from artefact, signal from noise. Experiments are often directed at taking up the challenge of designing better experiments: how can we learn more, and do it faster? How can we more cleverly circumvent flaws and limitations?

To begin with, researchers wish to explore whether there is even something worth investigating. There may be no theory in place to flesh out, much less to test. The very domain in which any eventual theory might live may be unclear. Even when the goal is to fill out or test a theory, intermediate experimental inferences are needed to bridge the data-theory gaps. Experimental phenomena may concern effects that the theory does not even talk about – an important part of their power. Theories, even where we have them, do not tell us how to test them. Nor will it do to imagine that theories are tested by joining various auxiliary hypotheses, hypothetico-deductively. Even in the rare cases where scientists can do that, it would not be a profitable way to go.

The aims of experiment – both in day-to-day practice, and in the startling discoveries on the cutting edge – are most aptly characterized by the simple desire to find things out.

Stability and Stubbornness: Experimental Knowledge Remains

The second gloss on the slogan concerns the continuity and growth of experimental knowledge. Despite changes in theory, even in the face of the need to reinterpret the significance of experimental inferences, claims that pass a stringent 'test of experiment' are generally stable. Even if the goal of the experiment is to estimate the parameters of a theory, the crux of good experiments is that the inferred estimates may be detached reliably. They are (as experimental physicists say) 'clean estimates' not sullied by unknowns. The deflection of light parameter in experimental general relativity, for example, is a clean effect that any adequate theory of gravity will have to accommodate.

Experimental knowledge grows, as do the tools for acquiring it, whether instrumentation, manipulation, computation, self-correction and so on. Continuity at the level of experimental knowledge points to a crucial kind of progress that is overlooked when measures of growth are sought in terms of large-scale theory change, in updating probability assignments to theories, or by means of other favoured macro-methodology schemes.

Independent Warrant for Inference

This leads to the deepest point about the 'own life' slogan: namely, the independent justification of experimental data and inference. Independent of what? The answer generally given is: independent of theory or at least 'high level' theory, inviting the criticism that theory always enters. But that mistakes what the 'own life' achievement is all about. What really matters is attaining freedom from whatever could be a threat to what the researchers are trying to find out. The interesting thesis is that experimental evidence and inference need not be theory-laden in any way that invalidates their various roles in grounding experimental arguments. They need not be dragged down by whatever is thus far unknown. While granting that experimental data are not given unproblematically, the position is that an experimental inference may be vouchsafed apart from threats of error. A particular experimental inference may be in error, but so long as errors are sufficiently understood and controlled we may discover, and avoid being misled by, them.

Some philosophers of experiment stress the independent grounding afforded by knowledge of *instruments*; others stress the weight of certain experimental activities such as *manipulation*. But neither is necessary or sufficient for experimental learning. Knowledge of instruments and astute manipulation may be important in obtaining experimental knowledge, but any means of reliable experimental learning may do as well or better, including computation, simulation, and statistical methods and modelling.

The term 'reliable' is notoriously ambiguous. I will put it to one side for now except to note that it does not suffice that the experiment will get it right in the long run, with high probability, asymptotically, or the like. A reliable experiment has to be capable of *controlling misinterpretations in the case at hand*, or at least within the time of a typical inquiry, research effort or report. The goal is not error avoidance but error control, which may be had by deliberately capitalizing on ways we know we can be wrong.

Canonical Errors: Terminology

In speaking of 'errors', I am not referring merely to *observational* errors, systematic or unsystematic, but rather mistaken *understandings of any aspect of a phenomenon*. Some may see this as a nonstandard use of 'error'. The ways that a claim or hypothesis H may be 'in error' include erroneous claims about underlying causes and mistaken understandings of any testable aspect of a phenomenon of interest. I am not drawing distinctions between experimental and theoretical, as some might. I am prepared to call a context 'experimental' (whether literal manipulation is present or not) insofar as error-probing capacities can be controlled and assessed.[2]

Co-opting a term from (frequentist) statistics, where the probabilities of methods for discerning errors are called *error probabilities*, I refer to an *error-statistical* approach (though the account is certainly not limited to formal statistical experiments).

While there are myriad types of mistakes in inference, I propose that there are a handful of error types and strategies for checking, avoiding and learning from them. I term these 'canonical' errors:

- mistaking chance effects or spurious correlations for genuine correlations or regularities
- mistakes about a quantity or value of a parameter
- mistakes about a causal factor
- mistakes about the assumptions of the data (for the experimental inference)
- mistakes in linking experimental inferences and subsequent claims or theories.[3]

While these are not exclusive (for example, checking assumptions and causal factors may take the form of testing for parameter values couched in models), each seems to correspond to distinct types of arguments and standards of evidence. Even so, I want to emphasize that these are just ways of categorizing strategies for far more context-dependent queries. There is a corresponding localization of what one is required to control, as well as what one is entitled to infer.

The interest is in capturing, quite generally, erroneous inferences: erroneous interpretations of data or erroneous understandings of phenomena and whatever can hinder the discovery of them. In one sense the goal is deeper and more interesting than getting a true hypothesis or theory. Even if, say, experimental relativists knew in 1930 that general relativity was true, they could not be said to have 'correctly understood' (relativistic) gravity. That demanded probing how gravity behaves in specially designed experiments.

Paying deliberate attention to errors, I claim, is at the heart of getting correct inferences and warranted interpretations of data. (So, clearly, the topic of this forum is near and dear to me.) Even granting that all models are wrong, and all theories strictly false, we are not prevented from getting a correct understanding of experimental stabilities and effects: *they have their own life!*

When Is It Bad for Data to Depend on Theory?

The most serious problem with theory-dependent inferences arises when the very hypothesis or theory under test is implicitly assumed. If it is predetermined that the interpretation of data is constrained to be in accord with a hypothesis H, whether or not H is correct, then the fact that data accord with H scarcely counts in its favour. This is to use a method that has no way of teaching us where H may be wrong. Enterprises that regularly proceed in this way would be considered pseudoscientific: data are not being taken seriously in the least. Even if the data set x accords with or 'fits' H, the procedure has maximal error probability: H 'passes' a test with minimal stringency or severity:

> If data x is generated by a method or procedure with little or no capability of finding (uncovering, admitting) the falsity of H, even if H is false, then x is poor evidence for H.

However, it may not be at all obvious, at least in interesting cases, that a method is guilty of such prejudgement. Whether data-dependent inferences (data mining, non-novel data) lead to high error probabilities needs to be scrutinized case by case.

For instance, many think that the fact that a hypothesis H was constructed to fit or accommodate the data, prevents the data also from counting as a good test of, or warranted evidence for, H. This 'no double-counting' requirement captures a general type of prohibition against data mining, hunting for significance, tuning on the signal and ad hoc hypotheses, in favour of requiring predesignated hypotheses and novel predictions. Whether (and when) inferences should be discounted, if not disallowed, when double counting has occurred has long been the source of disagreement and debate in philosophical and statistical literature. It is well known that if one is allowed to search through several factors and selectively report only those that show (apparently) impressive correlations, then there is a high probability of erroneously inferring that a correlation is real. However, it is equally clear that there are procedures for using data both to identify and test hypotheses that would only rarely output false hypotheses: the use of a DNA match to identify a criminal, radiointerferometry data to estimate the deflection of light, and entirely homely examples such as using a scale to measure my weight. Here, although the inferences (about the criminal, the deflection effect, my weight) were constructed to fit the data, they were deliberately constrained to reflect what is correct, at least approximately.

The secret to distinguishing cases, I argue, is staying clear about what error one needs to worry about, and whether the severity, stringency or probativeness of the test to avoid the error is compromised. What matters is not whether H was deliberately constructed to accommodate data x, but how well the data, together with background information, rules out ways in which an inference to H can be in error.[4]

Severity Plus Informativeness: Experimental Kuru

Merely avoiding error is scarcely sufficient for finding things out. True, before inferring H we want the experiment to have had a fairly high capability of unearthing flaws in H (stringency, severity), but aside from stringency we want informativeness. A key experimental aim is to carry out inquiries that will enable us to find something out, to extend our knowledge. The test or inquiry should have a fairly good chance of teaching us something. Or else we could stick to uttering tautologies and making unrisky claims. We are prepared to risk error because we want to learn. We deliberately construct experimental phenomena in order to learn about naturally occurring phenomena, but in such a way that error is controlled.

Consider the example of a brain disorder known as kuru (derived from 'to shake'), which ravaged the Fore people of New Guinea during the 1950s and 1960s. In its clinical stage, which lasts an average of twelve months, its sufferers go from having difficulty walking, to laughing uncontrollably, to being unable to swallow, to dying. Kuru and (what we now know to be) related diseases like BSE (mad cow disease), Creutzfeldt–Jakob disease and scrapie are 'spongiform' diseases – they cause the brain to appear spongy. (They are also called TSEs: transmissible spongiform encephalopathies.)

D. Carleton Gajdusek probed the disease experimentally without a clue about what a satisfactory theory of kuru might be.[5] He asked: *what causes kuru? Is it transmitted through genetics? Infection? How can it be controlled or eradicated?* No theory is needed to begin to ask how hypothetical answers to these questions might be wrong.

There was considerable data in the 1950s that kuru clustered within families, in particular among women and their children, or elderly parents. Might it then be genetic, like Alzheimer's disease, which produces similar plaques in the brain? But that was soon ruled out as erroneous, if only because it would have killed off the tribe. Researchers began to suspect that kuru was transmitted through mortuary cannibalism to maternal kin, a practice that both honoured the dead and let women in on the limited meat supply. Ending the cannibalism all but eradicated what had been an epidemic disease. This alone did not show the correctness of the mortuary cannibalism hypothesis. (Many of the tribe remained convinced it was witchcraft.)

In order to start experimenting, they had to dream up a method that would allow delimiting and probing possible causes through deliberately triggered experimental versions of kuru: *could kuru be transmitted to animals if their brains were inoculated with infected tissue from kuru victims?*

The similarities between kuru and scrapie in sheep led Gajdusek to begin experiments that would test whether kuru could be transmitted to animals.[6] Many of the attempts to transmit kuru failed, but given the difficulty of the process, the length of transmission and much else, the failures showed only the inadequacy of the experiments: they were not evidence of the falsity of the kuru hypothesis *H*. Nor were they uninformative; they were the basis for improved experiments. Years of lengthy experiments would show that kuru could be experimentally transmitted to chimpanzees, monkeys and other animals if they were injected with specially prepared infected brain extracts (from kuru victims). They studied experimental kuru to explore actual kuru.

Experimenters did not pretend to have gathered anything more than evidence about kuru's transmission; they called it a 'slow virus' (given the lengthy incubation), understood as little more than whatever it was they were transmitting. A report of errors not yet probed was an important part of the research report. They had at most a family of hypotheses about which to inquire. But researchers were excited by the fact that posing questions about deliberately triggered experimental phenomena had opened up platforms for engaging with a variety of TSEs. What was being learned lived a life of its own apart from theories.

No one expected that what had been learned about kuru would have revolutionary implications for a novel type of infectious particle (with no nucleic acid) that Stanley Prusiner[7] would term a *prion* (proteinaceous infectious particle).

Arguing From Error and Revolutionary Science

That experimental knowledge may get a life of its own is clearly relevant to avoiding a problem that gives philosophers of science so much trouble: theory-laden data. Kuhn contrasted revolutionary science with so-called normal science, wherein researchers flesh out theories and solve articulated problems or 'puzzles' posed by the aims and methods of the large-scale theory or 'paradigm'. While few philosophers nowadays accept the full Kuhnian (or even the Lakatosian) picture, it seems they refuse to shake off one or more of its stumbling blocks. Many think that any anomaly can be avoided by changing enough of the background; while logically true, such a tactic is a poor way of learning.

A well-known Kuhnian position is that while paradigms provide the researcher with tools for conducting science within a paradigm, they supply no tools for breaking out of that very paradigm: 'Paradigms are not corrigible by normal science at all'.[8] Not so. Local experimental results may show unavoidable anomalies that thereby overthrow hard cores with the same methods at hand. In one sense, I am simply taking seriously Kuhn's idea that 'severity of test-criteria is just one side of the coin whose other face is a puzzle-solving (i.e. a normal science) tradition'.[9] Where we differ, or seem to, is that I deny that there is a break when it comes to correcting theories. Genuine experimental knowledge is stable, and any future theory of the phenomenon, to be adequate, must accommodate it.[10]

Overthrowing the Central Dogma of Biology

Reproducible data began to accumulate, indicating that whatever was causing kuru could not be eradicated with techniques known to kill viruses and bacteria; furthermore, victims were not showing the presence of antibodies that would be produced by infectious elements that possessed nucleic acid. At the same time, researchers observed that infectivity (of both kuru and scrapie) was weakened by factors known to modify proteins.

As the experimenters described in detail, prions, whatever they may be, resist inactivation by UV irradiation, boiling, hospital disinfectants, hydrogen peroxide, iodophors, peracetic acid, chaotropes and much else. So if it were a mistake to construe kuru as having no nucleic acid, then at least one of these known agents would have eradicated it. A general pattern of argument emerges.

Experimental Argument from Error

We argue that H is a correct construal of data x when the procedure would have unearthed or signalled the misinterpretation, but instead regularly produces results in accordance with H. This is an example of what I call an *argument from error.*[11] (Note: data *x* would generally be a vector of outcomes possibly from several sub-experiments.)

They did not know what this non-virus was, and Prusiner said as much ('the transmission of experimental Kuru became well established, but its mode of action remained puzzling'). The genius of experiments is precisely in their allowing us to ask a single question at a time: Nucleic acid or not? Protein or not? Whatever it was, it appeared to be an infectious substance that was neither a virus nor bacteria, at odds with 'the central dogma' of biology. There was a genuine experimental phenomenon that would not go away, as well as a platform for launching ever-deeper probes of error.

It's All in the Planning (and Design) of Experiments

Experiment living 'its own life', circumventing obstacles to finding things out, is a consequence of something that has received too little attention in philosophies of experiment: experimental design and planning. I include under this rubric any aspect of deliberate planning for the collection, 'treatment', modelling and analysing of data. Experiments are distinguished from passive observation precisely because of the role of deliberate design in delimiting and controlling factors that would otherwise interfere with learning about effects of interest. Even with 'fortuitous' observations, researchers may try to mimic what experiment offers.

This brings up the question of what kinds of examples philosophers of experiment might fruitfully consider in constructing an experimental account. If perfect controls were attainable then we would not be in a very illuminating domain of learning, so I am interested in more challenging kinds of cases. For the same reason, we can learn the most by looking beyond cases where the primary goals are safely limited to a particular data-generating mechanism. It is precisely for this reason that I am often led to considering cases where the best one can do is to model experiments statistically. Statistically framed inquiries, by being explicit about the fact that the primary question is being 'embedded' into a statistical data-generating mechanism (DGM), emphasizes the need to connect back 'up' from what is learned about the DGM to the substantive or primary question of interest.

Homes for Experimental Life

Experiments have lives of their own, but it should be a real life, not a life in the street, with its own parameters, models, theories. An account that begins with given statements of evidence and hypotheses will not be relevant to the actual practice of experimental science. This recognition has been one of the most important ways that experimental philosophers have revealed the shortcomings of accounts that seek evidential-relation 'logics' assumed to hold between any given statements of data and hypotheses. That an adequate account of experiment should explicitly incorporate the methods and models of obtaining and analysing data is by now well accepted.

What about the work that goes into designing (or specifying) hypotheses to test or infer? Much less has been said about this and it is a central gap which I encourage philosophers of experiment to fill. Leaving issues of 'discovery' and specification vague is one of the reasons for the lack of clarity on the issue of when data-dependent specifications and selection effects matter. It has also led philosophers of induction and evidence to overlook the crucial value of very local, piecemeal hypotheses 'on the way' toward arriving at substantive claims and models.

I want to highlight a triad of components in planning, running and interpreting experiments in practice, including:

- questions, hypotheses
- data collection
- data analysis and interpretation.

There may be only a vague question or a loose family of speculative hypotheses (pre-data), and these typically will be very different from hypotheses or inferences arrived at (post-data). To capture a single unit of experimentation (perhaps as big as a given inquiry or published paper) with plenty of roomy niches for these components (see Figure 3.1):

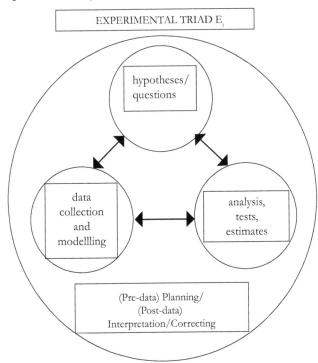

Figure 3.1: Experimental triad.

However, the three components are actually intimately connected, and the planning concerns all three at once. What is inferred at one stage may well be data for the next. Moreover, the deliberate use of raw data, differently modelled, is crucial to the problem of testing assumptions for a given experiment. The details by which experiments can self-correct their own assumptions should be a central component of any adequate account of experimental learning.

C. S. Peirce: *Quasi-Experiments*

It is perhaps unsurprising to find support for this conception in the work of Charles Peirce, who promoted the idea that the key to inductive inference was the ability to self-correct and learn from error. Peirce, himself an experimental scientist, coined the term 'quasi-experimentation' to cover the entire process of searching, generating and analysing the data, and using them to test a hypothesis and 'this whole proceeding I term induction'.[12]

Further, for Peirce, the 'true and worthy' task of *logic* was to 'tell you how to proceed to form a plan of experimentation'.[13] Were we to take this Peircean idea seriously, we would get beyond sterile *a priori* logics of evidence to something much more interesting and relevant.

Appraising/Ensuring Error-Probing Capacities

The key question directing experimental planning is how to split off questions that can be probed in terms of experimental data-generating procedures that will afford adequate control of interfering factors. The main considerations are these:

> (1) to ensure (pre-data) that experimental analysis of the data will be possible so that something is very likely to be learned, even if the experiment is botched.
> (2) to design 'custom-made' effects that might not even exist in nature in order to make the data talk to us (about aspects of their origins).

Design specifications alter the error-probing capacities of experiments, but far from sullying the results, they are the key to an objective interpretation of results. In statistics we use characteristics of the experimental test to determine the probability that a discrepancy from a test hypothesis will be inferred erroneously (type 1 error), and that discrepancies will be erroneously overlooked (type 2 error).[14] The analogous critique is conducted informally in scrutinizing experiments.

Experimental Accounts Should be Objective While Empirical

Even with this more complex picture of experimental ingredients, we need not forgo the philosopher's yen to identify overarching patterns of argument. The argument patterns follow variations on the basic structure of learning from error. The nitty-gritty details involved in cashing out such arguments – organized by a handful of error types – call for empirical experimental knowledge. Yet the account itself is normative.

Given especially the consciousness raising afforded by the focus on experiment, it is odd still to hear some philosophers maintain that if we relinquish the idea of a logical relationship between statements of given evidence and hypotheses (evidential-relation logics), then we also give up on objectivity. Achinstein,[15] for instance, declares that philosophical accounts of evidence are irrelevant to scientists because they are *a priori* while scientists evaluate evidence empirically. I think this is a serious mistake.

Popper recognized that there was a tension between our intuitions about stringent tests that avoid being ad hoc, and the aim of constructing formal logics of evidence, but fell into the same false dilemma. While Popper famously spoke of corroborating claims that passed sincere attempts to falsify, he was at a loss to make operational 'sincere attempts to falsify', although he clearly did not intend it to be a matter of psychology. Merely 'trying' to falsify, even if we could measure effort, would be completely beside the point: one must deliver the goods. Clearly, the Popperian idea that claims should be accepted only after passing 'severe attempts to falsify them' is in the error statistical spirit; but Popper never had an account of evidence that could do justice to this insight.[16] His followers, the critical rationalists, have not gone further except to note that we need to go beyond formal logics to recognize what Lakatos and Musgrave call the 'historical' nature of evidence.[17]

I heartily agree that we need to take into account how hypotheses are constructed and selected for testing. But too often what one counts as evidence is thought to be subjective, relative, historical or psychological. There are equivocations here of which we must steer clear: different stages of inquiry, just like different regulatory bodies (e.g. OSHA versus the EPA in the USA), may use different standards, but given the chosen standard, whether or not it is met need not itself be subjective, historical or psychological. That there is a multitude of different, and context-dependent, ways to satisfy requirements of experimental evidence does not imply that judging whether evidence is good or poor is a relative matter.[18]

In order to appraise purported claims of evidence, without which the context is not properly experimental, one must scrutinize the (error-statistical) properties of experimental tools and methods, their capacity or incapacity to distinguish errors and discriminate effects. Combining tests strengthens and fortifies the needed analysis. This is an empirical task all right, but it can be free of debilitating assumptions, which is at the heart of the 'own life' achievement.

How Experiments Free Themselves

An experiment need not actually exclude interfering factors to be 'free' of them. It suffices to *estimate (or simulate)* their influence, 'knock out' or *subtract* out effects, or cleverly *distinguish* or disentangle the effects of different factors. This gives rise to a variety of strategies, notably:

Amplifying effects:

An excellent way of learning from error is to deliberately magnify it. A pattern may readily be gleaned from noisy data by introducing known standards and measuring deviations or residuals. By magnifying distortions, mere whispers can be made to speak volumes. Through simulations and statistical modelling we may find out what it would be like if a hypothesized construal of data were wrong, or if a given factor were operating.

Self-correcting and/or showing robustness:

It is often still said that an experimental inference is only as reliable as the sub-experiments involved; but with good experiments, garbage in need not be garbage out. The overall inference may be far more reliable than any individual sub-experiment. Experiments achieve *lift-off.* Violating background assumptions might render them less efficient, but the cornerstone of good experiments is to allow the inference still to stand (robustness). By contrast, if satisfying assumptions of data and experimental models are highly sensitive to minor or common errors in attainable data, the experimental inferences have a hard time getting free of them. Fortunately, techniques for discovering and even exploiting such obstacles can be deliberately introduced into shrewdly crafted experiments.

Appealing to repertoires of error:

Clearly, any particular experimental triad E_i uses and builds on a general background of information which I dub a repertoire of errors. This would include errors that have and have not been ruled out in the area of interest, as well as pitfalls and success stories in dealing with the type of error that may be of concern. An adequate philosophical account of experiment should include such repertoires; there is considerable latitude as to how this may be accomplished. Historical cases may illuminate instances where errors were made several times before they were canonized. New ideas for cross-checking could well emerge.[19]

If we have put together a sufficiently potent arsenal for unearthing types of mistakes, we may construct a testing procedure that is highly capable of revealing the presence of a specific error, if it exists, but not otherwise. We may call this a *severe error probe*. If no error is uncovered with a demonstrably severe error probe, there is evidence it is absent. By contrast, if no error is discerned, but the procedure was fairly likely to miss it, then we are not warranted in inferring it is absent, at least on the basis of this particular probe. (Finding no evidence of error is not evidence of no error.)

Finding such inseverity is itself highly informative: properly used, it offers one of the most effective ways to invent new hypotheses. In particular, it supplies ideas for hypotheses we would not have been able to discriminate with existing tests. We quite literally learn from error – more precisely, from recognizing the limited error discriminating abilities of a given test.

Experiments Can 'Use' but Still Not Depend on Theories

Some have complained that experimental philosophers 'were throwing the baby out with the bathwater' in their emphasis on low-level experimental hypotheses.[20] We overlook, they say, how we invariably use background theories. What they overlook is how experiment may 'use' theories, or make use of theoretical knowledge, while not being threatened by them. They have their own life!

There is no problem in 'relying' on background theories

- to hypothetically draw out their consequences for testing, or
- when they have passed severe tests of their own, or
- when the only aspects being relied on are known to hold sufficiently.

Finally, an experiment can free the interpretation from assumptions that might introduce error (theoretical or other); namely, state them in the sum-up. (This is akin to a conditional proof in logic.)

Some might find it unusual for a philosophical account of evidence to require explicit consideration of both what has and has not been adequately learned from a given experiment; but that is what I am proposing. It will not appeal to neatniks! But the complexity pays off in providing an account that does justice to the cleverness of experimental practice, and at the same time enables long-standing philosophical problems of evidence to be resolved.

Agnosticism on Scientific Realism

Underlying the 'throwing out the baby with the bathwater' remark, in some cases, is the assumption that anyone who does not set out to build an account of theory acceptance is an anti-realist and limited to 'empirical adequacy' or a mere 'heuristic' use of theories. Musgrave wonders why I am prepared to accept experimental claims but not theoretical ones; but as noted at the outset: I never made any such distinction.[21] The models linking data to hypotheses and theories all contain a combination of observational and theoretical quantities and factors, as I see them. Since the same issues of warranting experimental knowledge arise for realists (of various stripes) and non-realists, the entire issue of scientific realism is one about which the error statistician may remain agnostic.

Experiment is Piecemeal: Threats of Alternative Hypotheses Squashed

Being able to implement these strategies leads to 'getting small', to specifying a question that will restrict or control erroneous interpretations of the kinds of data we are actually in a position to collect. This circumvents the familiar problem of 'alternative hypothesis objections'.

A typical challenge is: how can you rule out ways H can be false when there are always members in the 'catchall hypothesis' – including claims not even considered? To determine the likelihood of data under the catch-all hypothesis, as Wesley Salmon puts it, 'we would have to predict the future course of the history of science'.[22] The catch-all is indeed problematic for accounts that require listing all alternatives to a hypothesis H, or all the ways H might be false. But experiment deliberately delimits and reduces the factors that could be responsible for observed effects. With effective experimental design and data generation, many of the logically possible explanations of results may be effectively rendered actually or virtually impossible. If a certain gene is successfully knocked out of a mouse, it is not responsible for the effect; if the researchers do not know who got a placebo, then preconceptions of effectiveness are removed or at least diminished. The goal of experimental design is to specify a question that will restrict or control erroneous interpretations of the kinds of data we are actually in a position to collect. We can exhaust the space with respect to that one question. Even if we err, we may arrange things so that there is a good chance it will be detected in subsequent checks or attempts to replicate.

Rather than trying to distinguish a hypothesis or theory from its rivals, experiment sets out to distinguish and rule out a specific erroneous interpretation of the data from *this* experiment. In so doing the experimenter is not restricted by initial questions posed; a post-data appraisal is needed. Given the data, experimenters ask: what, if anything, do these data enable ruling out?

One could rationally reconstruct experimental inquiry using models of large-scale theory change, of Bayesian updating, or of decision theory with specified losses – once an episode is neat and tidy. The ease of doing so, some think, is one of the weaknesses of such reconstructions. Like a paint-by-number algorithm for the Mona Lisa, they do not capture how the learning took place (or the painting was created). They are backward looking, not forward looking, and fail to do justice to actual experiments. In setting sail to find things out, pre-data specifications of an exhaustive set of rival substantive hypotheses are atypical; much less do we have, or want to interpret the data in light of, cost or benefit functions as a decision-theoretic construal requires. Even evidence-based policy (or other subsequent decisions) should rest on a valid experimental knowledge base.[23] Certain low-level claims (about parameters, directions of effect, observed correlations) may appear overly simple if they are thought to be the final object of study, but for exhausting the space of a local error (e.g. $\mu \leq \mu'$ or $\mu > \mu'$) they are just the ticket!

Arguing from Coincidence

The powerful form of argument that experiments provide is often described as an 'argument from coincidence': there is no way that all of these well-known instruments and independent manipulations could consistently produce certain effects or concordant data, were they all artefacts of instruments. It is not merely that the concordant results are formally improbable. As Hacking notes, to suppose that they are all instrumental artefacts is akin to invoking a Cartesian demon of the instruments.[24] This inference to a non-artefact (or 'real effect') is an instance of my *general argument from error*: 'We argue that there is evidence that (the artefact's) error is absent when a procedure that would have (with high probability) unearthed the (artefact's) error fails to do so, but instead consistently produces results indicating its absence'.

This can also be put in terms of inferring a genuine, non-chance effect.[25] We may deliberately create artefacts and discern how readily they are revealed. This teaches us about the error-probing capacities of our overall experiment, which combines many results. This is all part of the background repertoire of knowledge of errors.

Knowing that an effect is 'real', in the sense of non-artefactual, however, is one of the weakest kinds of experimental knowledge. Still it is important. It was the first error on my list. It may arise to check whether an instrument is working, or even to ascertain whether there is any real effect worth exploring. Experimenters want to know if they would be wasting their time trying to explain effects that could readily be accounted for 'by chance'.

Arguing from Coincidence to What?

By leaving arguments from coincidence at a vague level, however, they are often appealed to as warranting much more than they actually allow. Avoiding the error that needs avoiding in order to infer that there is evidence of a real (not spurious) effect does not directly warrant hypothesized explanations of the effect. (That is why statistical significance is not substantive significance.)

Yet many say they do: would it not be a preposterous coincidence if all these different experiments $E_1, E_2, E_3, E_4 \ldots E_n$, yielded data x in agreement with theory or hypothesis H, if H were false? Would it be? That is what we would need to figure out. If it would, then the argument shows H to be well or severely tested in my sense. If it would not, we should not be accepting all of H on the basis of data x.

To suppose we should, or that scientists do, is to take a pattern of argument that works where we can exhaust, and rule out, all the possibilities and apply it in general. Instead I say we should put our epistemology of experiment at the level of experiment. Such local experiments offer standard or canonical ways of exhausting answers, e.g. nucleic acid or not? Far from seeking to infer all of H before its time, the engine for experimental knowledge grows through understanding why and when we would not be warranted in doing so.

Relevant Variability Depends on the Error We Need to Rule Out

Most importantly, the varied experiments E_1, E_2, E_3, E_4 ... E_n, must be shown to be relevantly varied! Errors should have ramifications in at least some of the other experimental trials. Whatever might threaten one experiment must not also be able to be responsible for an error in the others. If all witnesses have been bought off, a 'wide variety' of them may not yield anything more probative than one; if all the samples are contaminated, their agreement does not help.

If our experimental account is to be forward looking, as I urge, then the focus should be on how to move, in the constrained fashion that experiments allow, from what is known at a given stage to learning more. These may be baby steps, but at least they will be taken securely. Transmission of kuru was known to be real in the 1980s, say, but do the data warrant a 'protein-only' explanation? No. And experimenters could not even experimentally probe such a question before learning how to construct testable forms of the 'protein-only' hypotheses. So let us turn to some aspects of experimental learning about prions. I obviously cannot here relate this rich and decades-long episode; my goal is to uncover a few of the gems it offers for experimental learning from error.

An Experimental Platform for Understanding Prion Diseases

Key aspects of what enables experiments to live lives of their own are exemplified by the use of 'animal models' in probing the transmission of kuru and other TSEs (in this case mostly in mice and hamsters): the models supply a general platform on which to probe various aspects of a phenomenon of interest (here, prion transmission) so that what is learned remains regardless of subsequent reinterpretations or overarching theories still to emerge.

As the researchers got better at purifying prions, it became clear that the minimum molecular weight necessary for infectivity was so small that it excluded viruses and any other known infectious agent.[26] Prions were found to contain a single protein dubbed PrP. To the researchers' surprise, PrP was found in normal animals – so it does not always cause disease. They dubbed the non-pathogenic, common form PrP-C; the pathogenic, or scrapie form, PrP-Sc. Then they deliberately designed experiments to discover if they 'had made a terrible mistake' (and prions had nothing to do with it).[27] They had not. One way of learning how matters would have gone had they been wrong, however, was to create transgenic mice with PrP deliberately knocked out. No such knock-out mice were able to be infected with PrP-Sc.

Getting Good at Learning about Prion Transmission

But how is it that different patterns of infection were observed, with different incubation times and apparent species barriers? The answer seems to point to different 'strains' of prions, but strains are the sort of thing only viruses were thought to have.

The researchers did not know what they would find, and an adequate (forward-looking) account of experiment should be at home with this fact of experimental life. What they can and did do is design experiments that would give them a good chance of learning something about species barriers to infectivity.

Transgenic mice with a hamster PrP gene were created: when inoculated with mouse prions, they made more mouse prions; when they were inoculated with hamster prions, more hamster prions. They were onto something that might explain the species barrier. Mice are not normally infected with hamster prions, but hybrid mice, created with portions of both hamster and mouse protein sequences, they discovered, could be infected with either mouse or hamster proteins! By combining transgenic approaches and computer modelling methods, they were able to produce mice susceptible or resistant to prion disease in predictable ways. This was the basis for experimental learning.

Decades before they had even begun to understand the mechanism behind the observed species barriers in infection, prion transmission between species was modelled as a stochastic process (based on scrapie in sheep). During the 1980s, experimental manipulation enabled them to predict and control transmission, so that 'it becomes a nonstochastic process'. They knew their understanding of transmission was growing when they could reduce the incubation period from 600 days to 90 with continual injections of pathogenic brain tissue. (Understanding different patterns of species transmission is obviously relevant to the question of BSE/mad cow disease in humans – when and why humans are infected, and how to detect it.)

For biologists, yeast is a great model organism: it has a growth cycle of only eighty minutes. But, as is typical, damping down one error heightens another. Only by connecting the transgenic and normal animal models, the yeast and the hamster, and so on, was it possible for one experiment to serve as a relevant check on the others, permitting the overall argument to be free of threats to reliable learning.

The Only Correct Interpretation of the Data: It is in the Folding

An important mode of learning from error is to consider why certain experiments are incapable of telling experimenters what they want to know; and this case offers several illustrations. By mixing synthetic versions of the two proteins together in a test tube, they were able to convert common prions (PrP-C) into scrapie prions (PrP-Sc) (in vitro), but they did not understand what was actually causing the switch. Moreover, they knew they did not, and they knew something about why. The infectious form has the same amino acid sequence as the normal type: studying the amino acid sequence does not enable us to reveal what made the difference. If exceptions to the 'central dogma' were precluded, and it is assumed that only nucleic acid directs replication of pathogens, there was no other place to look. But what if transmission by pathogenic proteins occurs in a different way?

Maybe the difference is in the way the protein is folded. Researchers hypothesized that the scrapie protein propagates itself by getting normal prions to flip from their usual shape to a scrapie shape. This would explain 'transmission' without nucleic acid, and sure enough they were able to replicate such flipping in vitro. But understanding the mechanism of pathological folding required knowing something about the structures of common as opposed to scrapie prions. A big experimental obstacle was not being able to discern the prion's three-dimensional structure at the atomic level. Exploiting the obstacle provided the key.

Magic Angle Spinning: Exploiting an Obstacle

The central difference between normal and pathogenic prions permits the normal but not the abnormal prion to have its structure discerned by known techniques, e.g. nuclear magnetic resonance (NMR) for solutions: The normal form, PrP-C, is soluble; PrP-Sc is not.

NMR spectroscopy provides an image of molecular structure: Put a material inside a very high magnetic field, hit it with targeted radio waves, and its particles react to reveal their structure. But it will not work for clumpy scrapie prions, PrP-Sc. Maybe solid-state NMR could detect them?

Even so, they would need trillions of molecules to get a signal – amplification – but this would also amplify the interference of neighbouring molecules in the non-soluble PrP-Sc. They want to find out *what it would be like* if they were able to make it soluble, even though they cannot literally do so. They need to amplify to get a signal, but also somehow subtract the interference of neighbouring molecules. Here is where 'magic angle spinning' enters.

The Magic is to erase the influence of these neighbouring molecules.

If the sample, crushed into a powder form, spins within a magnetic field at a special angle to that field – 54.7 degrees – the influence of a molecule's environment is cancelled out. The effect on the spectrum from the magnetic interactions between the molecules vanishes.[28] Knowledge of the magic angle stems from quantum mechanics, but that it works to negate interactions is shown with known molecular structures.

While the molecules cannot all be lined up at the magic angle, they can, *on average*, if they are spun fast enough (with respect to every other molecule in the sample). In this we have one of the deepest and least appreciated aspects of the role of models of relative frequencies, and intermediate statistical models, in linking data to questions about phenomena of interest. We want to know the molecular structure that is actually there, and we can control the errors that threaten such learning by translating the problem into one that requires only the detection of averages.

In practice, this often means the sample needs to spin at a rate of 25,000 to 50,000 revolutions per second (in order to get all the molecules in the sample to 'sing in same key'). It can still take weeks of listening.

Building up arguments along these lines, researchers could argue from error that the transmission of prion disease is due to protein misfolding. Prusiner's particular 'prion only' theory (for which he won a Nobel prize in 1997) is that prions target normal PrP and turn it into a form that folds and clumps, ultimately causing brain cells to rupture. The abnormal prion moves on to normal prions, pinning and flattening their spirals, akin to what one researcher describes as a deadly Virginia reel in the brain.[29]

As is typical with powerful experimental tools, we have very general instruments (e.g. solid state NMR) that enable us to check if we have gotten it right for known samples. By deliberately turning the NMR turbine too slowly for the magic angle, we get expected distortions, showing we are interacting as predicted. We argue from error that we correctly understand an instrument when we know how to get it to produce the patterns of distortions expected from deliberate violations. This is supplemented with entirely distinct instruments for further cross-checking.

The known distortions and limits of each experimental analysis are key to linking experimental knowledge to the phenomenon of interest. Generalizing into a much wider spectrum of diseases involving pathological folding (e.g. Alzheimer's) simultaneously generalizes our knowledge as it affords more stringent, interconnected experimental probes of error.

Concluding Remarks

A model of experimental inquiry (a triad making up a data-experiment-hypothesis grouping, Figure 3.1) enables questions of interest to be asked in relation to some aspect of the proposed data-generating procedure. Inquiry is successful to the extent that it is free of threats from obstacles to finding things out, that is, *by getting a life of its own.*

The experimental framework I propose allows flexibility to move within as well as continually add to the three main components. It provides a general platform that is not domain-specific, enabling standard checks to interlink individual experimental inferences, known to be limited, partial and distorting. It offers concrete homes within which to pinpoint how an inference from raw data to a primary hypothesis of interest may fail to be warranted:

- the experimental inference that *is* licensed fails to provide evidence for the primary hypothesis
- the experimental hypothesis has not passed a severe test
- the assumptions of the experiment are not met sufficiently by the actual data.

With transgenic mice designed to produce tons of normal prions, with synthetic prions, with protein folding in vitro, and so on, experimenters create a series of experiments $E_1, E_2, E_3, E_4 \ldots E_n$, that enable arguments from error to stand. The immediate goal in each is not to rule out rival theories, but to rule out a mistaken interpretation of results from this experiment.

But what do the pieces say about actual (real-world) prion disease? Might these be relevant only for Frankenstein mice, yeast, synthetic prions and mimicked scrapie in a test tube? To combine these pieces requires understanding the errors or distortions that remain, and those being avoided or subtracted out. (What's being amplified? What's being silenced?) We can investigate and learn about these limits and distortions by deliberately amplifying and controlling them in known or canonical cases.

In so doing we capitalize on experimental benchmarks,[30] calibration standards,[31] a variety of extrapolation models and a suite of relevant repertoires of errors. By the time magic angle spinning is used to learn about the structure of prions, it is a well-understood and reliable instrument. Imagining that an instrument works only when it is used on a known sample is to imagine that it can read our minds, and that it conspires to trick us just when we are faced with unknown samples. This would be like my claiming that all the scales have conspired to show that I have gained weight when I have not, while allowing that the scales work fine with objects of known weight. If any of the scales were faulty, this would show up when they were used on objects of known weight. We need only deny that a mysterious power conspires to make all observations fit our deductions on known samples, but not on unknown ones. That would be a radical obstacle to learning![32]

The powdered PrP-Sc spun around in magic angle spinning turbines can tell us about the three-dimensional structure of PrP-Sc because we know how the turbines work with known solid-state specimens. While only the structural relationship is discerned, that is all we need for this piece of the puzzle. Other interlocking hypotheses make the pieces relevant to what we want to know. To understand specifically how and why this works we should consider how each experiment checks, unearths, amplifies and erases the threats or shortcomings of others.

Even today, the prion evidence can be accommodated by those who still maintain there could be some nucleic acid locked within the prion protein; indeed, they revise and develop their alternatives to deliberately account for the known evidence. It was only after decades of accumulated growth of experimental knowledge that defenders of the older paradigm could construct alternatives such as virions (a virus wrapped in a protein) that could explain why no protein is discerned, despite years of trying. Nonetheless, the idea that transmission is through protein misfolding is granted even by the alternative virion theories, as I understand them. As such, work on developing rivals does not really hamper progress (though it frustrates 'prion-only' theorists); and the corroboration of the general prion transmission principle is not compromised. In general, the development of rivals to a hypothesis *H* that could account for existing data plays an important role in experimental learning, even where it must be denied that the data are evidence *for* the rivals constructed. It is a way for defenders of *H* to uncover their own errors, and for sceptics about *H* to show which aspects have so far not been well-tested. With well-grounded experimental knowledge, these rivals do not alter the well-testedness of the aspects severely probed: they have their own lives.

Scientists are rarely fully explicit about or even aware of why their methods and strategies work when they do. Providing such an illumination would be an important task for a future philosophy of experiment: the key is to unearth how experiments manage to free themselves from threats of error.

4 MODELLING MEASUREMENT: ERROR AND UNCERTAINTY

Luca Mari[1] and Alessandro Giordani

Introduction

The relation between error and uncertainty in measurement is a complex subject, in which philosophical assumptions and modelling methods are variously intertwined. Both the concepts 'error' and 'uncertainty' aim at keeping into account the experimental evidence that even the best measurements are not able to convey definitive and complete information on the quantity intended to be measured. The pragmatic approach prevailing in the literature on instrumentation and measurement today often leaves the distinction between error and uncertainty implicit, as it were just a lexical issue in which a single, stable concept is expressed differently.[2] While at the operative level this interchangeability can find some justification, e.g. the rule customarily adopted for the analytical propagation of errors and uncertainties is indeed the same, whether such concepts are alternative or complementary is a particularly important topic, also because of the claimed current 'change in the treatment' of the subject, 'from an Error Approach (sometimes called Traditional Approach) to an Uncertainty Approach.'[3]

In this chapter, after having introduced the problem and presented the general background of our analysis in the context of the realist versus instrumentalist opposition in philosophy of science (pp. 80–2), we propose a solution by exploiting the concept of models as mediators between the informational level of the propositions and the ontological level of the world (pp. 82–4). In particular, it is shown how the concept of truth can be suitably applied to propositions characterizing the relation between a quantity value and a *quantity construed as an entity specified in the context of given model*, while the concept of (un)certainty suitably applies to characterize the relation between such a modelled quantity and a *quantity as an empirical entity specified by reference to a concrete object*. This is the basis for an in-depth analysis of the complex connection between truth and (un)certainty and the development of a framework in which these concepts are combined.

After a brief revision of the operational contribution given by the *Guide to the Expression of Uncertainty in Measurement* ('GUM')[4] (pp. 84–7), a simple model of the operation performed by a measuring instrument is proposed, and then compared with the two other basic processes in which the instrument is involved: calibration and metrological characterization (pp. 87–94). The analysis of the three processes highlights that a concept of (operative) true value, and therefore of error, can be introduced by pushing the position that measurement is able to convey '(operative) pure data'. By relaxing some of the given hypotheses, it is finally shown (pp. 94–6) that in a broader and more realistic context, measurement could be understood as a knowledge-based process, where both (operative) truth and (un)certainty play a significant role.

Two Perspectives on Error and Uncertainty

The relation between error and uncertainty in measurement is so deeply rooted in foundational issues that even the basic related terminology is somehow controversial and has to be preliminarily agreed upon. Still, even a trivial case, such as the measurement of the length of a common object such as a table, suits the purpose of exemplifying the required concepts. We will call:

- the table a, the *object under measurement* (it could be a phenomenon, an event, a process, and so forth);
- the length Q, a *general quantity* (sometimes called attribute, observable, property, kind of quantity, undetermined magnitude);
- the length of the table q_a, *an individual quantity* (sometimes called magnitude), which is here the quantity intended to be measured, the *measurand* for short;[5]
- an entity v such as 2.34 m, a *quantity value*.

The basic objective of measurement m can be then described as the assignment of the best (in a sense to be specified) quantity value to a given measurand, i.e. to a general quantity considered of the object under measurement.

$$q \xrightarrow{\;\;m\;\;} v$$

The theoretical assumptions concerning the measurand are crucial for the relation between error and uncertainty in measurement. In particular, in a hypothetical spectrum of theoretical options about the knowledge of the quantitative characters of the world, two extremes can be identified: a *realist* standpoint, according to which knowledge depends on how the world is structured in itself, and an *instrumentalist* standpoint, according to which knowledge depends on how the world is determined by our conceptual structures. This alternative significantly affects the interpretation of the nature of measurement and the involved entities (see Table 4.1).

Table 4.1: The realist and the instrumentalist standpoints

According to the realist standpoint:[6]	According to the instrumentalist standpoint:[7]
both general and individual quantities exist independently of measurement	*neither* general nor individual quantities exist independently of measurement
individual quantities are *actually* related by numerical ratios	individual quantities can be *operationally* related by numerical ratios
once an individual quantity is *selected* as a unit, all other individual quantities of the same general quantity are *determined* by a number	once an individual quantity is *introduced* as a unit, all other individual quantities of the same general quantity can be *represented* by a number
measurement is a process aimed at *discovering* the measurand, where the quantity value states the result of such a discovery	measurement is a process aimed at *assigning* a quantity value to the measurand

Let us suppose that an experimental situation is given where repeated measurements, which are assumed to be performed on the same or similar objects and with respect to the same quantity under specified conditions, provide different quantity values. The two standpoints interpret the situation as follows.

According to the realist standpoint, since quantities exist in the world, an object a is characterized with respect to a general quantity Q by a definite individual quantity q_a, to which a true value $v(q_a)$ is in principle associated. The fact that repeated measurements provide different values for q_a is justified in terms of measurement errors, which make knowing the true value impossible *in practice*. The assumption of the existence of the true value $v(q_a)$ leads to the idea that such a value can be approximated in increasingly improving ways as the measurement process is enhanced and the effects of measurement errors are reduced.

According to the instrumentalist standpoint, since quantities do not exist in the world, a is not characterized with respect to Q by any definite individual quantity q_a. The fact that repeated measurements provide different values for the measurand is accounted for in terms of an uncertainty which makes knowing true values impossible *in practice and in principle*. The assumption of the non-existence of the true value $v(q_a)$ leads to the idea that measurement uncertainty is an irreducible feature of measurement and characterizes 'the dispersion of the quantity values being attributed to the measurand, based on the information used'.[8]

In the context of measurement science, the concept of *true quantity value* is an unavoidable element in the first perspective, as witnessed by its foundational role in the classical theory of errors, and an element of no significance in the second one.

This tension has led to a critical attitude towards the concept of true quantity value, a concept which is now used (if outside academic papers and books it is even used at all) in a way that always stresses that it is an ideal, unknowable element. With this acknowledgment, the VIM3 has proposed the above-mentioned idea of two 'approaches', the 'Error Approach' and the 'Uncertainty Approach', contrasted with each other with respect to the expected objective of measurement: in the Error Approach, to determine an estimate of the true value of the measurand which is as close as possible to that unique true value; in the Uncertainty Approach, to decide an interval of values which can justifiably be assigned to the measurand. Hence, this opposition is further interpreted as related to the alternative: unique true quantity values versus intervals of quantity values, where sometimes, possibly to maintain a reference to the tradition, the hybrid concept 'intervals of true quantity values' is adopted.[9] The change in the article – *the* true value versus *a* true value – might be taken as a lexical symbol of this conundrum.

All these elements support the hypothesis that error and uncertainty are incompatible concepts, which cannot be reconciled even at the operative level. As a consequence, the option for the concept of uncertainty seems to result in giving up the role of truth, and therefore error, in measurement. Is this actually the case? Our claim here is that this incompatibility of 'approaches' is inconsistent and unjustified, and that it can and should be overcome.

Framing the Perspectives

The realist versus instrumentalist opposition on the role of error and uncertainty in measurement can be inscribed in a broader contraposition concerning knowledge as such,[10] which can be roughly presented as follows. The realist standpoint emphasizes that the conceptual framework we use to describe the world can be abstracted from the observable world and provides us with information about the structure of the world itself. The instrumentalist standpoint, on the other hand, emphasizes that such a conceptual framework is constructed by ourselves and then used to interpret the world in such a way that the interpretation determines the reference of the concepts. These perspectives are both, in different ways, related to the problem of the truth. According to the realist standpoint, which assumes the building blocks of the propositions to be obtained from the world, any proposition has a definite *truth value*, depending on its actual correspondence with a considered 'portion of the world'. On the contrary, the instrumentalist standpoint states that the building blocks of the propositions are introduced by us, so that a proposition has only an *aptness value*, expressing its capability of allowing us to successfully interact with the world.

Some recent developments in philosophy of science[11] have suggested a way to conciliate these perspectives by focusing on the role of *models* in the knowledge of the world, and on the function of measurement, a crucial knowledge tool, accordingly.[12] Indeed, in the last decades the view of models as mediators for understanding and interacting with the world has been variously stressed: the current picture, while too complex and controversial to be accounted for here, can be outlined in its general traits as follows:[13]

- any portion of the world of interest is conceptualized by means of a model, in which the pertinent objects and properties characterizing that portion as a system under consideration are identified;
- this model is used both as a theoretical tool for interpreting our concepts and as an operational tool for studying the corresponding portion of the world.
- this way of conceiving the relation between models and portions of the world permits us to dismiss both (1) the position according to which propositions allowing for a successful interaction with the world are true with respect to the world and (2) the position according to which propositions are never true, but only more or less successful.

The proposal here is that a *proposition can be true with respect to a model* and a *model is more or less similar to the world*, and as such it is more or less successful. Indeed, let us assume that a portion of the world can be conceptualized by means of a model and that a model can be described by means of propositions. In addition, let us assume that

- to be successful is a property that can be significantly ascribed only to instruments, and particularly to models, which are non-linguistic entities,

whereas

- to be true is a property that can be significantly ascribed only to propositions, i.e. to specific linguistic entities.

Then a proposition is neither more or less successful nor true with respect to the world, but it can be a *true* description of a *successful* model, and thus *true with respect to that very model*.

By interpreting measurement in terms of models, the concepts of true quantity value and measurement uncertainty can be combined without being forced to interpret uncertainty in terms of doubt about the truth of a measured value. The basic idea is that truth refers to the attribution of a quantity value to a modelled quantity, while uncertainty characterizes both such attribution and the degree of similarity between the model of that quantity and the actual, empirical quantity, and therefore also to their combination, i.e. the attribution of a quantity value to the quantity.

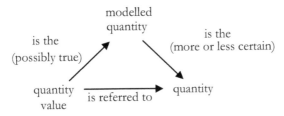

Hence, the attribution of a quantity value to a quantity involves uncertainty both because the quantity is modelled in such a way that it is difficult, or even impossible, in principle to know its value (e.g. because it is modelled as its value is in a set of real numbers) and because the model of the quantity is not necessarily accurate (e.g. because influence quantities are not properly taken into account).

A simple, but paradigmatic, example comes from the way in which the problem of measuring the area of a paper sheet is solved. A twofold idealization is customarily introduced, according to which the object under measurement is modelled as a rectangle and the measurand is defined as a general quantity, area, of that object, which can be evaluated by positive real numbers and which is not influenced by other quantities. The solution is then obtained by measuring the length of the two sides of the sheet and taking their product. Hence, in this model the object under measurement has an individual area, whose true value is obtained by multiplying the lengths of its sides. Still, since such a model is an idealization of the actual sheet, which is surely not 'perfectly rectangular', to convey correct information about the measurand it is necessary to estimate the possible discrepancy between the value obtained by virtue of the ideal measuring procedure and the value, indeed the set of values, obtained by virtue of the actual measuring procedure, and then to express this discrepancy in terms of uncertainty.

This conclusion shows that the realist and the instrumentalist standpoints are not required to be thought of as fundamentally opposed. Still, this position, despite its interest, constitutes only a partial solution to the problem of the relations between error and uncertainty in measurement, and this is for two reasons: first, because the recent developments of the operational perspective in metrology have introduced new motivations for casting doubts about the concept of a unique true value; second, because the analysis can be deepened to show that in the very case of measurement the concept of true value has application with reference not only to modelled quantities, but also to actual quantities, i.e. to the world.

Focusing on the Operational Perspective

A landmark in the development of a widespread adoption of uncertainty modelling in measurement has been the publication, in 1993, of the GUM, aimed at establishing 'general rules for evaluating and expressing uncertainty',[14] building on a recommendation issued in 1980 by a Working Group on the Statement of Uncertainties convened by the Bureau International des Poids et Mesures (BIPM) and approved in 1981 by the Comité International des Poids et Mesures (CIPM), the world's highest authority in metrology. To this purpose a concept of measurement uncertainty was to be preliminarily adopted: innovating the tradition, the GUM focuses on operational, instead of epistemological or ontological, issues, and stresses in particular:

1. the construal of the concept of measurement uncertainty;
2. the introduction of the concept of definitional (there called 'intrinsic') uncertainty;
3. the elimination of the concept of true quantity value as identical with the concept of quantity value.

Regarding point (1), measurement uncertainty is interpreted as uncertainty on the quantity value to be attributed to the measurand, where indeed the GUM recommends a measurement result to be a pair (measured quantity value, measurement uncertainty), from which the usual interval form, $v \pm \Delta v$, can be obtained.[15] Hence, measurement uncertainty becomes a parameter 'characterizing the dispersion of the quantity values being attributed to a measurand',[16] instead of a parameter estimating the distance of the quantity value from the true one.

> The definition of uncertainty of measurement … is an operational one that focuses on the measurement result and its evaluated uncertainty. However, it is not inconsistent with other concepts of uncertainty of measurement, such as (i) a measure of the possible error in the estimated value of the measurand as provided by the result of a measurement; (ii) an estimate characterizing the range of values within which the true value of a measurand lies … Although these two traditional concepts are valid as ideals, they focus on unknowable quantities: the 'error' of the result of a measurement and the 'true value' of the measurand (in contrast to its estimated value), respectively. Nevertheless, whichever concept of uncertainty is adopted, an uncertainty component is always evaluated using the same data and related information.[17]

The classical distinction between random and systematic errors – which attributes being random or systematic as a feature of error itself – is set aside, and a new classification, 'Type A' versus 'Type B', is adopted, which, most importantly, refers not to the sources of uncertainty, but 'to the way in which numerical values are estimated', 'by statistical methods' or 'by other means' respectively. This switch is claimed to give an operative solution to a long-standing problem: 'random and systematic errors … have to be treated differently [and] no rule can be derived on how they combine to form the total error of any given measurement result'. Hence, the concept of error can be avoided here and the philosophical assumptions concerning truth and true values discharged: there is no necessity of making a reference to a supposed true value, since there is no possibility of evaluating the distance between the estimated value and the true value. Rather, measurement uncertainty is acknowledged as deriving from several possible sources (and the GUM lists several of them: (a) incomplete definition of the measurand; (b) imperfect realization of the definition of the measurand; (c) nonrepresentative sampling – the sample measured may not represent the defined measurand; (d) inadequate knowledge of the effects of environmental conditions on the measurement or imperfect measurement of environmental conditions; (e) personal bias in reading analogue instruments; (f) finite instrument resolution or discrimination threshold; (g) inexact values of measurement standards and reference materials; (h) inexact values of constants and other parameters obtained from external sources and used in the data-reduction algorithm; (i) approximations and assumptions incorporated in the measurement method and procedure; (j) variations in repeated observations of the measurand under apparently identical conditions) and nevertheless all of them are to be formalized as *standard uncertainties*, i.e. standard deviations. This assumption leads to a single rule for combining all components of uncertainty – the so-called *law of propagation of uncertainty* – which is thus applicable to components obtained by both Type A and Type B evaluations.

Regarding point (2), above, the fact that the information obtained by measurement is supposed to be related to a measurand introduces an unavoidable interpretive component in the measurement problem, which then requires the measurand to be specified. This implies giving a description of both the object under measurement and the environment expected when the measurement takes place. Hence, in principle a measurand cannot be completely described without an infinite amount of information. The consequence is that the measurand, now defined as the quantity *intended* to be measured,[18] is to be distinguished from the quantity actually subject to measurement, and a definitional uncertainty remains as far as this distinction is concerned.[19]

Regarding point (3), above, the GUM states that the word 'true' in 'true value' is 'unnecessary', because 'the "true" value of a quantity is simply the value of the quantity'.

These three points highlight some significant tensions in the GUM approach. The concept of definitional uncertainty is introduced, but then deprived of any operative import, since the definitional uncertainty is simply assumed to be negligible with respect to the other components of measurement uncertainty. Under this hypothesis, the value attributed to the measurand can be assumed as 'essentially unique' (and note: 'essentially', not e.g. 'practically'), so that the expressions 'value of the measurand' and 'true value of the measurand' are taken as equivalent. On the other hand, precisely this position, where the equation between the 'true value' of the measurand and 'the best estimate of the value' of the measurand is assumed, leads to the problem of building a conceptually consistent framework. It is surprising to declare both that the true value is eliminable, since it is unknowable, and identical to the best estimate of the value, which is evidently known. Alternatively, it is surprising to declare both that the true value is the value of the measurand, identical with the best estimate of the value of the measurand, and that this value is not truly representing the measurand. Thus, stating that 'true' is redundant seems here to be just a lexical position, which leaves unmodified the problem about true values.

In addition, there is a further price to be paid for this emphasis on the operational side of the problem. The assumed unknowability of true values transfers to another pivotal concept for measurement, the one of measurement accuracy, defined as 'closeness of agreement between a measured quantity value and a true quantity value of a measurand'.[20] As a consequence, then, 'the concept "measurement accuracy" is not a quantity and is not given a numerical quantity value', plausibly an elliptical way to state that were measurement accuracy considered a quantity then its value would be unknowable in its turn because of the reference in its definition to a true value. Still, accuracy is customarily listed among the features of measuring instruments, and a numerical value for it is indicated. The way outs which are sometimes adopted to solve the puzzle, e.g. redefining the concept relating not to a true value but to a 'conventional true' value (is then truth assumed to be conventional?) or a (generic) reference value,[21] do not really help us to cope with the general issue.

As a synthesis, while the introduction of the definitional uncertainty, which leads us to improve our initial picture, where a model true value is introduced, can be viewed as a positive contribution of the GUM, the apparently deflationist strategy underlying the elimination of the concept of true value,[22] instead of clarifying the frame, seems to obscure important characters of the measurement process and to neglect the fact that not all quantity values are unknowable, as, e.g. nominal quantity values – which represent design constraints – are known by specification.

Rethinking the Concepts and their Relations

With the aim of further exploring the concepts of measurement error and measurement uncertainty, let us introduce an admittedly simplified model of the empirical core of a measurement process, *a measuring instrument*, interpreted as a basic device enabling quantity representation. The VIM3, which defines it as a 'device used for making measurements, alone or in conjunction with one or more supplementary devices', notes that 'a measuring instrument may be an indicating measuring instrument or a material measure'.[23] An indicating measuring instrument (material measures can be omitted in the following discussion) is a 'measuring instrument providing an output signal carrying information about the value of the quantity being measured',[24] such an output signal being called the indication.[25] An indicating measuring instrument operates, in fact, as a generic transducer, which dynamically produces an output quantity q'_{out} as the effect of its interaction with (an object characterized in particular by) an input quantity q_{in}.

A simple example of a transducer which can be operated as an indicating measuring instrument is a spring: in response to the application of a force (the input quantity q_{in}), the spring stretches and a length is thus obtained (the output quantity q'_{out}).[26] What makes a spring a measuring instrument? There are three basic conditions.

MC1: the output quantity q'_{out} is assumed to reliably provide information on the quantity being measured q_{in}.

This is a condition of predictable input–output behaviour, requiring that (i) the mapping $q_{in} \to q'_{out}$ can be formalized as a function τ_q by an underlying theory, or at least a black box causal modelling of such behaviour, and that (ii) the transducer which implements τ_q is properly constructed and operated according to a given procedure. The function τ_q corresponds to the *empirical component* of measurement.

$$q_{in} \xrightarrow{\ \tau_q\ } q'_{out}$$

MC2: the output quantity value v' corresponding to q'_{out} is assumed to be known.

This condition prevents a never-ending recursive process: were the output quantity subject to measurement in its turn, a further transducer would be required (a measuring instrument whose input quantity is length in the case of a spring), and for its output quantity this condition should apply. Hence, it is required that (i) the mapping $q'_{out} \to v'$ can be formalized as a function d_{out}, and that (ii) such function is known. The function d_{out} corresponds to the *evidential component* of measurement.

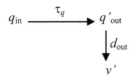

MC3: an input quantity value v corresponding to the output quantity value v' can be obtained.

This results from the hypotheses that (i) the function τ_q between quantities is mirrored by an invertible function τ_v between quantity values, and that (ii) such function is known. The application of τ_v^{-1} constitutes the core of the *inferential component* of measurement.

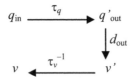

Whenever conditions MC1–MC3 are satisfied, measurement can be performed, as formalized as a function m obtained by the composition of the empirical, the evidential and the inferential components, $m = \tau_v^{-1} \bullet d_{out} \bullet \tau_q$.

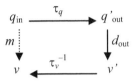

As a synthesis, this model of the measurement process, even though extremely simple, shows that several components have to be coordinated to establish the relation between a measurand and a measurand value.

Indications and Indication Values (Discussion on MC2)

MC2 *is crucial for the present discussion.* Let us analyse it a little bit more thoroughly. First of all, the relation between quantities and quantity values is clearly a fundamental topic for measurement, which is indeed aimed at associating a measured quantity value to a quantity, the measurand.[27] On the other hand, there are some clues that such relation is still open for discussion and better clarification.[28]

Returning to transducers, and therefore to measuring instruments: are indications then quantities or quantity values? If the lexical confusion is not justified, in this case the conceptual superposition is acceptable, precisely because of MC2 (in the following discussion we will remove the ambiguity by using 'indication' for a quantity and 'indication value' for a quantity value). In any measuring instrument there must be a point in which the relation between a quantity and a quantity value is assumed as given, as a *pure datum* in an operative sense, i.e. as an entity that, although possibly further specifiable, can be recognized and recorded without any additional interpretive or inferential process. According to MC2 this point is where indication values are produced, thus justifying the term 'evidential', introduced above, for this component. In fact, measuring instruments are designed so as to make the mapping from indications to indication values straightforward, being typically implemented as a process of pattern recognition, performed by human beings or technological devices: the observation of coincidence of marks, the classification of an electric quantity to a quantized level to which a digital code is associated, the numbering of right answers of a test, and so on.[29] All of them are assumed to be unproblematic processes leading to values for the given quantity, i.e. the instrument indication, under the assumption that a pattern recognition of sufficiently low specificity is both truthful and its outcome cannot be further refined in the context. In this way, the information conveyed by the mapping from indications to indication values is the best one which can be achieved by means of the instrument. Accordingly, the obtained indication value can be properly said to be the true quantity value of the corresponding indication. Of course, this is a revisable, operative truth, so that the term *operative true quantity value* could be adopted to denote this entity, which has to be systematically distinguished from the *model true quantity value* introduced before. On the other hand, such operative true values are indication values, not measurand ones (length values instead of force values in the case of a spring). Hence, such truth is still not sufficient for measurement.

Transduction and Calibration (Discussion on MC3)

Since the measurement function m is obtained as $\tau_v^{-1} \bullet d_{out} \bullet \tau_q$, a critical problem in ensuring the correctness of m concerns the way in which the mapping τ_v^{-1} from indication values to measurand values is obtained. Such a problem is solved by means of instrument *calibration*. Interestingly, the description of the instrument operation for measurement and calibration is the same: the instrument interacts with an object, and an indication is obtained as the result of the transduction of an input quantity of the object. Still in a simplified model, the conditions for calibration are as follows.

CC1: a set of reference objects ref_j is available, each of them providing an input quantity $q_{ref,j}$ of the transducer and thus producing a corresponding output quantity $q'_{out,j}$

$$q_{ref,j} \xrightarrow{\quad \tau_q \quad} q'_{out,j}$$

CC2: the quantity value v_j for each of such reference quantities $q_{ref,j}$ is assumed to be known.

$$
\begin{array}{c}
q_{ref,j} \xrightarrow{\quad \tau_q \quad} q'_{out,j} \\
\downarrow {\scriptstyle d_{ref}} \\
v_j
\end{array}
$$

In this case, too, a never-ending recursive process must be prevented: were the reference quantity subject to measurement in its turn, a measuring instrument would be required, and the problem would arise again with respect to that instrument. Hence, it is required that (i) the mapping $q_{ref,j} \to v_j$ can be formalized as a function d_{ref} and that (ii) such function is known. The function d_{ref} corresponds to the *evidential component* of calibration.

Together with MC2, CC1 and CC2 ensure that information on both $q_{ref,j}$ and $q'_{out,j}$ is provided.

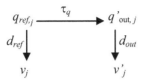

Calibration is then aimed at producing the map $v'_j = \tau_v(v_j)$, as a set of pairs, $\{<v_1, v'_1>, <v_2, v'_2>, ...\}$, or an analytical synthesis of them, for example resulting from the hypothesis of linearity of τ_v, so that $v'_j = \alpha + \beta v_j$ for given parameters α and β.

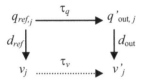

This highlights the fundamental role of calibration for the inferential part of measurement, which is based on the relation encoded in τ_v. In its turn, the construction of τ_v is based on the evidence concerning the available operative *pure data*, i.e. about indication values, as given by d_{out}, and about reference values, as given by d_{ref}.

References and Reference Values (Discussion of CC2)

CC2 *is the second crucial element in this discussion.* Like MC2, it supposes the operative availability of unproblematic values, assigned by convention, or through a chain of responsibility delegation, typically guaranteed by a calibration hierarchy (sometimes called a traceability chain) from a primary measurement standard. Together with MC2, it ensures the possibility of mirroring the empirical mapping τ_q, which connects quantities, i.e. empirical entities, with the informational mapping τ_v, which connects quantity values, i.e. informational entities. It is in this way that in a calibrated measuring instrument *the empirical component is reliably linked to the evaluation component.*

This point is delicate. In some presentations the whole problem of measurement is introduced by assuming that the input of the measuring system is the true measurand *value*,[30] that experimental errors 'hide' in the mapping to indication values, with the consequence that error theory/analysis has the purpose of estimating such input value despite the superposed errors. This is an overly simplified, and actually misleading, position: being empirical devices, measuring instruments interact with (quantities of) objects, not values. In particular, a value for the measurand is the final outcome of the process, not its starting point (but under the above-mentioned realist hypothesis that 'numbers are in the world'). Any criticism of true values based on this assumption is thus well founded but, as we are going to argue, this does not imply that truth, and therefore error, must vanish from the scope of measurement.

Measurement

As has been considered in the previous section, in both calibration and measurement the empirical mapping τ_q is operated. The difference lies in the interpretation and the purpose of this operation:

- in calibration the mapping τ_v is unknown and is looked for on the basis of the knowledge of the values v_j associated to references and of indication values v'_j;
- in measurement the input quantity value, v, is unknown and is looked for on the basis of the knowledge of τ_v and of the indication value v' obtained by transducing q_{in}.

Measurement exploits calibration information by inverting it:

$$v = \tau_v^{-1}(v')$$

Hence, the measurement m of the measurand q_{in} is performed by:[31]

1. empirically transducing q_{in} to an indication $q'_{out} = \tau_q(q_{in})$;
2. mapping q'_{out} to the corresponding value $v' = d_{out}(q'_{out})$;
3. mapping v' to the value $v = \tau_v^{-1}(v')$ attributed to the measurand.

In synthesis:

$$v = m(q_{in}) = \tau_v^{-1}(d_{out}(\tau_q(q_{in})))$$

According to this simplified model, the whole picture of *measurement* is as follows.

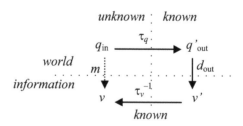

The condition on the basis of which measurement is possible is the construction of τ_v, as achieved in the process of *instrument calibration*, to which the following diagram applies.

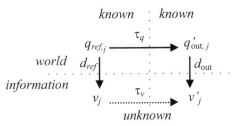

Metrological Characterization

Together with calibration and measurement, the empirical mapping τ_q is also exploited in a third kind of process: the metrological characterization of measuring instruments (the VIM3 calls it 'verification', 'provision of objective evidence that a given item fulfils specified requirements',[32] a similar process). Indeed, neither calibration nor measurement are aimed at producing information on the metrological capability of a measuring instrument. For this goal two basic processes (in particular) can be designed:

An input quantity q_{in} is repeatedly applied to the measuring instrument and a scale statistic (e.g. sample standard deviation, $s(.)$) on the sample of the produced indication values v'_j is computed:

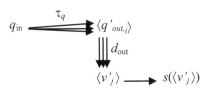

An input quantity q_{in} whose value v is assumed to be known is repeatedly applied to the calibrated measuring instrument, a location statistic (e.g., sample mean, $m(.)$) on the sample of the produced measured quantity values $\tau_v^{-1}(v'_j)$ is computed, and then v and the value of such statistic are compared:

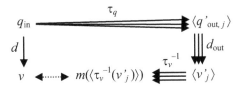

The first process does not require q_{in} to be provided by a measurement standard nor the measuring instrument to be calibrated, but is critically based on the stability of q_{in}. Its aim is to convey information on the stability of the mapping τ_q, a property usually called *measurement precision* (according to the VIM3 the 'closeness of agreement between indications or measured quantity values obtained by replicate measurements on the same or similar objects under specified conditions').[33] Under these assumptions, any non-null value for the given scale statistic has to be considered as the indicator of *errors* in the transduction behaviour of the measuring instrument.

The second process is more demanding, since it requires not only the stability of q_{in} but also the knowledge of its value v, as typically obtained by means of a measurement standard, together with the calibration of the measuring instrument. Its aim is to convey information on the stability of the mapping τ_v, and therefore of the calibration itself, a property called *measurement trueness* (according to the VIM3 the 'closeness of agreement between the average of an infinite number of replicate measured quantity values and a reference quantity value').[34] Under these assumptions, any difference between v and the location statistic has to be considered as the indicator of *errors* introduced by the fact that the calibration information is not correct anymore.[35]

It is fundamental to note here that:

- the mentioned errors refer to a measuring instrument but are identified and evaluated in a process which is not a measurement; only under a (customarily reasonable, in fact) hypothesis of stability of the measuring instrument, this information about its erratic behaviour can be assumed to hold also for measurement;
- precision and trueness, and then accuracy, are features related to the measuring instrument behaviour, and more generally to measurement if the measurement procedure is taken into account, and not to measurement results, although the information they convey is appropriately exploited in the evaluation of measurement data to assign a measurement result.

Conclusions: Reshaping the Framework

The realist and the instrumentalist perspectives can be composed, at first glance, by allowing for model true values and interpreting uncertainty as representing the acknowledged discrepancy or dissimilarity between a model and the modelled portion of the world. Still, we have also highlighted that (i) it is possible to be uncertain as to the model to choose in order to analyse a given portion of the world, and that (ii) it is possible to admit true values with respect to indication values and reference values. Accordingly, the diagram

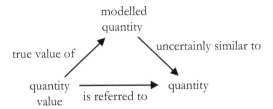

does not provide a sufficiently general account of the significance of the concepts of true value and uncertainty in measurement: some uncertainty affects the modelled quantity, and the concept of truth can be applied to values of some actual quantities too.

As a conclusion, the following synthesis, asking for further developments, can be offered.

A measurand can be modelled at different levels of specificity (in the sense according to which the measurand 'length of this metal rod at the temperature of 30°C' is more specific than 'length of this metal rod'). The chosen level of specificity is typically not the most specific one, so that there is an uncertainty as to which model, among the ones that specify the chosen model, would be the most similar one.

A connection between quantities and quantity values is the outcome of the measurement of measurands but is also assumed, as a definitional assignment, in the cases of indication quantities and reference quantities. Since such quantities are produced by a designed process, it appears to be legitimate to consider the values assigned to them as their operative true values.

In addition, if the simplified model *of* measurement introduced so far is embedded in a more realistic context, the role of models *in* measurement has to be taken into account. The idea is that the quality of measurement results is affected not only by measurement errors but also by other causes, which in the specific context of the given measurement are not empirically controllable and therefore can be evaluated only on the basis of given interpretive hypotheses. The resulting effects are expressed in terms of *uncertainty* in measurement results. Among such other causes there are the following ones (these are just hints: each of them could involve a much more thorough analysis).

Modelling the calibration hierarchy. The working standards exploited in instrument calibration customarily are calibrated in their turn, through a calibration hierarchy. While each step of this process is a calibration, the trueness-related errors might not combine linearly. Hence, instead of trying to conceive a complex super-model including the whole calibration hierarchy, the expert knowledge on the resulting effects can be exploited, and elicited in terms of the uncertainty of the quantity value of the working standard.

Modelling the transduction process: identifying the laws. The transduction behaviour of the measuring instrument is not perfectly characterized. If the transducer is assumed to obey a parametric law (typically: it is linear, i.e. the transducer sensitivity is constant in the measuring interval) its calibration is greatly simplified but at the price of an approximation, which can be taken into account in terms of uncertainty of the involved quantity values.

Modelling the transduction process: closing the system. The transducer exploited in measurement is not perfectly selective: the transduction process is perturbed by several influence quantities, so that its output depends not only on the stated input quantity. While in principle each influence quantity might be measured in its turn and its effects properly characterized and then eliminated, this process would lead to a never-ending recursive process, due to the fact that in the measurement of an influence quantity some influence quantities should be taken into account. Typically, some simplifying hypotheses are assumed on the effects of the influence quantities, expressed as an uncertainty of the indication value.

Modelling the measurand: identifying the quantity intended to be measured. It has to be admitted that the input quantity of the transducer might not be the measurand, i.e. that the quantity subject to measurement is not the quantity intended to be measured. In such a case a model has to be adopted to infer the measurand value from the available information and, of course, this model could be acknowledged as implying some simplifications, whose effects can be expressed as an uncertainty on the stated measurand value.

Modelling the measurand: defining the quantity intended to be measured. The measurand as such might be defined only up to a given approximation, resulting in a definitional uncertainty, 'the practical minimum measurement uncertainty achievable in any measurement of a given measurand', according to the VIM3. The metrological model of the measurand is the source of an important kind of uncertainty, which cannot be eliminated by means of experimental means. If it is impossible to define *the* quantity that is intended to be measured, then it is, even in principle, impossible to determine *its* quantity value.

These examples highlight that measurement uncertainty can be assumed as an encompassing concept by means of which the quality of measurement results is expressed by taking into account both the effects of measurement errors and the approximations due to measurement-related models. As a consequence, 'the Error Approach' and 'the Uncertainty Approach' to measurement are not only compatible but actually both required for an appropriate evaluation of measurement data: measurement errors are a component of the usually broader set of causes of measurement uncertainty.

5 HANDLING UNCERTAINTY IN ENVIRONMENTAL MODELS AT THE SCIENCE–POLICY–SOCIETY INTERFACES

M. Bruce Beck

Introduction

In support of a 1990 UK House of Lords private members' Bill – to enable construction of a barrage across an estuary in a politically high-profile part of northern England – a complex hydrodynamic model had been used to assess the distribution of pollution, should the barrage be built. In the run-up to the General Election of 1987, then Prime Minister Margaret Thatcher had stridden across the industrial wasteland alongside the River Tees, cast her arms open, inviting her television audience to take in the barren scene, and proclaimed 'We shall regenerate this'. The barrage, oddly enough, was to be the spark of that regeneration. The model had consistently forecast that pollution in the future would be shifted out to sea, far away from the inland barrage. While the opponents of the proposed barrage did not succeed in preventing passage of the Bill, they were granted the right to request a sensitivity analysis of the model. In the course of that analysis a substantial error was discovered. When corrected, pollution was forecast to accumulate against the barrage. The group responsible for the model and its forecasts was obliged to issue a formal public apology.

To be precise, the error was a programming error, rather than one of either the science underpinning the model or the values assigned to the parameters (coefficients) in its mathematical expressions. Even so, the systematic tinkering with the model that uncovered the error had to do with quite modest changes, of plus or minus a few per cent, to the value of a critical hydrodynamic parameter. The changes had to be modest, because that parameter was held to be 'universally known' by the model's authors. The implication was that this was a 'physically based' parameter: not quite the constant of gravitational acceleration, but of that ilk; hence it was in need of no further adjustment, such as that entailed in calibrating the model against field observations. Yet this parameter's value had actually been estimated from a regression-like curve fitted to the customary spread of data, from experimental work on another estuary. Arguably, therefore, it was *not* that well known.

Pragmatically, politically and socially – and 'socially' largely in the sense of the affairs of the scientific community – there was neither the time nor the will to expect any better treatment of such uncertainty in the model at this particular science–policy interface. The sensitivity analysis was of the rudimentary 'one-at-a-time' kind:[1] that of taking one model parameter, changing its value to, say, 1.05 times its nominal, default, or 'best' estimate, and observing the resulting changes in the forecast model outputs. We knew already then (in the 1980s) that there was much more to the uncertainty in a model, to its consequences, and its analysis.[2]

But different disciplines in the science of the environment, and different individuals within each discipline, react to uncertainty and model calibration in different ways, to differing extents, and sharply so.[3] Then, as in the passage of the Tees Barrage Bill, and some two decades on, as in the deliberations of the (US) National Research Council (NRC) on dealing with uncertainty in *Models in Environmental Regulatory Decision Making*,[4] the instinct of many was (and remains) this: 'If only we could have a model based solely on the laws of physics, then we would have models with known constants – and surely no need of model calibration'. The NRC book had the ambition of becoming an authoritative piece of work, if not a definitive statement on its subject matter.[5] Yet to accommodate the wide divergence of disciplinary perspectives on uncertainty and model calibration, a separate box of text had to be drafted. It was entitled 'To Calibrate or Not to Calibrate'. For those whose profession is the modelling of air quality, 'calibration is viewed as a practice that should be avoided at all costs'.[6] For some, none of all this matters anyway. There will always be those for whom models are an abomination, witness the book – *Useless Arithmetic: Why Environmental Scientists Can't Predict the Future* – by Pilkey and Pilkey-Jarvis.[7] Surely, the observed data should be allowed to 'speak for themselves', should they not?[8]

The central spine of this chapter, as with that of Smith and Petersen (this volume), has to do with what we shall refer to (in shorthand) as 'deep uncertainty'. The phrase can be understood in a rather intuitive fashion.[9] Kandlikar et al. qualified it as a matter of both 'uncertainty and ignorance', which is to say (for them), 'a profound lack of understanding and/or predictability'. Our concern in this chapter is with handling deep uncertainty in *models*, as opposed to uncertainty in science in general, at the science–policy–society interfaces. We begin, in the next section (pp. 101–7), by inspecting the nature of this deep (model) uncertainty from the perspective of science, as it points itself towards those interfaces. We shall introduce the accompanying shorthand of some algebraic notation, so that we can be more precise about the nature and ramifications of this deep uncertainty. The third section (pp. 108–17) treats the same – the nature and consequences of deep uncertainty, that is – from the perspective primarily of policy. It begins by introducing a dichotomy, between (let us call them) sound science analysis and deliberative problem solving, since this reveals a strategic difference in attitudes towards handling uncertainty in models, especially in the settings of policy formation and the law.[10] For the arguments we compose in this third section, 1990 might be described as an *annus mirabilis*. The books *Uncertainty and Quality in Science for Policy*,[11] *The Fifth Branch*[12] and *Cultural Theory, or a theory of plural rationalities*,[13] were all published in that year. This section draws from each – what determines pedigree in science for research (Funtowicz and Ravetz); the extension of peer review to science other than for research and discovery, i.e. science for regulation and policy (Jasanoff); and the paradoxically deep uncertainty of socially constructed, contradictory certainties (Thompson and others) – to provide a synthesis: an anatomy of the problems, so to speak. Our next task (in the fourth section, pp. 117–26) is to address the differences brought about when uncertainty in models, as opposed to the overarching issue of uncertainty in science (in general), is projected into debates at the science–policy–society interfaces (as likewise in Smith and Petersen, this volume). We do so with a primary focus on the social facet of the issue, in particular, within the community of scientists, based on personal, empirical experience of this over the past four decades.

Most of the chapter, therefore, is about composing a map of the challenges of deep uncertainty: circling around it, as it were, from the perspectives of science (pp. 101–7), policy (pp. 108–17) and, more loosely speaking, society (pp. 117–26).

The fifth section (pp. 126–33) is but a brief companion discussion of some contemporary, candidate problem-solving approaches, spread across what might be called the life cycle of a policy: in the process of generating foresight, regarding whether a policy is needed in the first place; in shaping, disputing, and negotiating the form of a policy, as it proceeds from something provisional to something final; and lastly, in changing policy once implemented, in 'real time'. At a deeper, technical level, this will require us to engage briefly with subjects such as the trustworthiness (or validity) of a model and structural change in the face of structural error and uncertainty in the model.

But first we must address the matter of what we understand to be a 'model', since this may be the cause of some considerable confusion.

The Nature of a Model

We use words differently in different disciplines, with consequently quite different meanings attaching to these words. For Morgan, in her book *The World in the Model*:

> Models in economics are still mostly pen-and-paper objects depicting some aspect of the economy in a schematic, miniaturized, simplified way.
> [...]
> [Models] must be small enough in scale for their manipulation to be manageable in order that they can be used to enquire – indirectly – into the workings of those aspects of the economy depicted.
> [...]
> These manipulatable objects are the practical starting point in economic research work: they are used for theorizing, provide hypotheses and design constraints for laboratory experiments, are an essential input into simulations, and form the basis for much statistical work.
> [...]
> That process: the historical and philosophical changes from reasoning with words to reasoning with models, is what this book [*The World in the Model*] is all about.[14]

In the domain of economics, then, and according to Morgan, we see that models are devices for 'reasoning' in our heads, about the way the world works. They are 'small', as she says, and from them may be derived 'input into simulations'.

Here, in this chapter, models in environmental science should generally be thought of as anything but small; they will be Morgan's simulations on a computer. It is indeed the norm in environmental science, including the atmospheric and climate sciences, for models to be cast in the form of nonlinear, differential equations and for the number (order) of those equations to be many (high). Others, no less complicated, will be recognizable as agent-based models (ABMs). In these, sets of agents – as proxies for the likes of individual people, institutions, businesses, and so on – are simulated as they interact with and move about the environment.

The predominant high-order, differential-equation models of environmental science will not be reasonably transparent, a quality presumably essential in Morgan's focus on the difference between reasoning in economics with (her notion of) models, as opposed to reasoning with words. The models of environmental science are impenetrable to many onlookers; their inner workings are difficult to scrutinize, including in the legal process.[15] And yet their high order, one might suppose, ought surely to make them less prone to the challenges of deep uncertainty. Viewed from the perspectives of policy and society, they should evoke trust, appearing to be just so 'complete'. How could they possibly have omitted something of relevance to the decision to be made? As some kind of 'truth-generating machine' or 'answer machine',[16] how could they possibly be wrong? On the other hand, they might induce suspicion: of the 'wool being pulled over our eyes'; of something mistrustfully or dangerously furtive going on in the 'black box' of the computer.

In fact, these large models are not free of problems from the perspective of science, not least in respect of the way uncertainty may be propagated through them.[17] Their opacity lies at the heart of some of the most intractable issues of handling uncertainty in models at the science–policy–society interfaces, as we shall see on pp. 126–33, in particular, with regard to coming to a judgement on the trustworthiness of a model, upon whose forecasts a given decision or policy is to be founded.

From the Perspective of Science

We take a model to be a more or less complex assembly of atomistic, constituent hypotheses about the way the world works. The mathematical and/or logical forms of expression of these hypotheses and the manner in which they are pinned together – akin to the scientific visualizations of complex biological molecules[18] – we call the model's structure M. M is populated with parameters α, i.e. it is a function of α, so that we can write cryptically $M\{\alpha\}$.

We resort to building and using models because we wish to understand the way a system behaves – national economy, atmosphere, city, human body – and to conduct experiments with that behaviour in the laboratory world of the computer, before committing imminent decisions and polices to actions in the real world. Such systems are subject to input stimuli and disturbances (u), which have an impact ultimately on other things of personal and policy interest to us, such as water quality against the barrage in the River Tees, i.e., output responses (y). M transcribes the ramifications of the causes u into the effects y.

In short, we have the triplet $[u, M\{\alpha\}, y]$, wherein, technically, everything is uncertain: u, M, α and y. How we obtain the model and use it, when likened to the problem solving in an elementary textbook on mathematics, tells us something about the origins, propagation and consequences of uncertainty for environmental policy formation. In the abstract, therefore, think simply of the problem 'Given $[u, y]$, find M!', or that of 'Given $[u, M]$, find y!', or there again, 'Given $[M, y]$, find u!'. In other words, given a relationship between three entities, where two are presumed known, use the relationship to solve for the third unknown.

Deep Uncertainty in the Model

In less abstract and vastly less simple terms, the problem of 'Given $[u, y]$, find M!' is known formally as system identification, or model identification. It entails computational and analytical operations referred to professionally as model calibration (fitting the model to the data), model parameter estimation, model structure identification, model verification, and model validation (or model evaluation, as environmental modellers would prefer;[19] Box 1).

For us, with our interest in deep uncertainty, it is particularly important to distinguish in the matter of system identification between, on the one hand, discovery (identification) of the structure of M and, on the other, estimation of the values of the parameters α within that so identified structure. Consider, therefore, the distinction between that which is {presumed known} in the model and its complement, the {acknowledged unknown}. Calibration of the model against the observed behaviour, i.e. against the data on $[u, y]$, presumes the structure of the model M is known *a priori*. It is beyond dispute; it is fixed and immutable. For calibration, the {presumed known} amounts to a sturdy, self-sufficient entity; the {acknowledged unknown} is a matter at the fringe – as though the entity is an engineering structure, with but a little random vibrational play in its joints, for which the tightening up of a few nuts (minor adjustments to the values of α) will suffice. The appropriate number of equations (of state) for the model are known, as also are the correct forms of the mathematical expressions appearing in each equation, and as too is how what happens in one equation is related to what happens in another. We should observe, however, that choosing the 'knowns' to be included in the model, and the manner of their mathematical expression, will oblige the builder of the model to make some (subjective) choices.

The presumption in model calibration is that only the uncertainty in the values of the model's parameters α needs modulating. Epistemic errors are *not* expected; entertained is solely the possibility of minor aleatory uncertainty in the {presumed known}. The correct and appropriate instrument of prediction (for the purposes of policy formation) has been assembled; just some fine-tuning – calibration of the instrument – is needed (see also Box 1 and Mari and Giordani, this volume).

In model structure identification, in sharp contrast, the recognition is that there can be significant error in the {presumed known}, i.e. there may be substantial flaws in those bits of the science base incorporated into the model. The structure of the model M is not necessarily correct. Indeed, it is generally some distance from that presumption, especially in respect of its characterization of the chemical, biological and ecological dimensions of the environment. The model's structure is insecure. When stressed to the limit, in trying to reconcile its behaviour with that observed of the real thing ($[u, y]$), it is prone to buckling, if not failure, in its parts – the constituent, member-hypotheses of the {presumed known} – possibly complete failure as a whole. Some of the reasons for the failures in the {presumed known} may, or may not, originate in the counter-intuitive, but tangible presence of some of the unknowns in the {acknowledged unknowns} (and beyond, into a world/space of Rumsfeldian unknowns,[20] for example). To express all this in shorthand, we may talk of model structure error and uncertainty (or MSEU). MSEU signals (cryptically) what we mean in this chapter as deep uncertainty in the model. It is also indicative of the second dimension of model unreliability in Smith and Petersen (this volume).

Under such deep uncertainty – of the possibilities of substantial flaws in the model's structure and gross omissions from it – no mere calibration-associated presumption of security in the structure of M is tenable. Working relentlessly and deliberately to reveal, diagnose, and rectify the errors and flaws in the bits of the knowledge base incorporated into the structure of an inevitably flawed and uncertain model M, is referred to as model structure identification. It is definitively an 'error-driven' process. It has to do with the growth of knowledge.[21] It fully deserves to be one of the grand challenges of environmental modelling[22] – no. 7, in fact.

There will be uncertainty in the model M before identification and there will be uncertainty in it thereafter, this latter being modulated according to the uncertainty in $[u, y]$ and all the distortions and struggles of reconciling the behaviour of the model with that observed of the real thing. The pattern of uncertainties reflected in M after identification, hence in the posterior α, can be thought of as a summary fingerprint of those struggles, for better, for worse, epistemic flaws, omissions, deep uncertainty, and all.[23] But now we might claim to 'know' M – including a quantification and mapping of its uncertainty – and we can turn to the other rudimentary forms of problem solving.

Box 1: Models, Measurements, and Measuring Instruments

The burden of this chapter rests on what is done with the model (M), once developed. That prior development of the model, therefore, is one step removed from the focus of the chapter. Measurements of the real thing – the data $[u, y]$ describing the observed behaviour of a portion of the environment, to which recourse is made in model development – are gathered in before solving for 'Given $[u, y]$ find M!'. Measurements, we might say, are accordingly two steps removed from what transpires at the interfaces among science, policy and society. Thus removed, it is too easy for policy-focused discussions with models to take the data for granted, in particular, the subtleties and mechanics of their acquisition (Mari and Giordani, this volume). For various reasons, the conjunction of models, measurements and measuring instruments needs our attention, no matter how briefly or superficially.

Models in the Instrument

As we have prefaced this chapter with reference to Morgan's *The World in the Model*, so too do Mari and Giordani (this volume) cite the works of Morgan in introducing a key feature of their treatment of error and uncertainty in measurement. Their discussion, however, is more closely aligned with Morgan's notion of 'models as mediators for understanding and interacting with the world' than is the present chapter. Mari and Giordani do not necessarily have in mind models as complex computational entities (M), as is decidedly the case here.

Somewhere, then, we – as enquirers into the nature of things – presume there exist quantities $[u', y']$ that are the 'true' values of the input and output variables ($[u, y]$) of the environmental system of interest. Following Mari and Giordani, who invoke the device of a model in order to reconcile realist and instrumentalist stances on truth, measurement, error, and uncertainty, we would say here that a model M^u (of the measuring instrument) transcribes input u' into that instrument into the instrument's output, the observation u. Likewise, a model M^y transcribes input y' into another (different) instrument into observation y. In general, therefore, it is possible for us to imagine the mechanics of the instrument behaving as a model, M^I. It transcribes what is sensed at the 'fingertip' of the instrument, as it touches reality, into a number; a number manipulable *not* in our heads (in our case), but in a computer program, in particular, a computer program for the model M at the center of the discussion at the interfaces among science, policy, and society.

Calibration of the measuring device (the instrument) obliges the fingertip of the instrument to make contact with a known u' (or y') that is known to generate a reference value (number) for u (or y). Given two or more such reference points of contact[24] for any variable to be measured, hence the two or more corresponding pairs of $[u', u]$ (or of $[y', y]$), the settings on the instrument M^u (or M^y) can be adjusted. The instrument is thereby calibrated for use in sensing any u (or y) over a pre-determined range (of values), which range is defined by the 'reference points of contact'. Over this range the instrument is judged to be a valid measuring device.

In the process, the instrument is not dis-assembled into its constituent parts, merely its parameters are finely tuned. Not unlike calibration of the computational model M, the construction of the instrument, i.e. its structure – the physical manifestation of its conceptual counterpart (M^I) – is fixed, beyond further dispute. If disassembly, reassembly, possibly redesign of the instrument were to take place, we would now recognize this as the analogue of model structure identification (from the discussion of the main body of this chapter). It is worth noting that the ranges of $[u, y]$ technically bound the range of behaviours of the environmental system for which our computational model (M) has been calibrated.[25]

Instruments in the Model

For decades, process control engineers have known that the dynamic behaviour of instruments, i.e. that sensing $y'(t)$ at time t is not revealed as a number $y(t + \Delta t)$ until some time Δt after t, can be an important consideration in using models to design automatic control systems. As an abstract system (if nothing else), the instrument has a pattern of input-output behaviour, just as the system it is observing. One should not be at all surprised by this. The machinery of measuring instruments can these days be massively complex.

For decades, the biochemical oxygen demand (BOD) has been the subject of mathematical models of water pollution (almost certainly it was a part of the model of the Tees estuary). Historically, the BOD *measurement* was carried out over a period of five days (hence the delay Δt) under laboratory conditions in a sealed vessel, at constant temperature, and in the absence of light. One can build a model of this *measurement* process (M^I).[26] In fact, once was the time 'Given $[u, y]$ find $M\{M^I\}$!' was formally contemplated as the only means of unscrambling the nature of M as it related to pollution of the River Cam in the UK.[27] In short, extracting an understanding of the behaviour of the system cannot necessarily be undertaken independently of unscrambling and disentangling the behaviour of the measurement device. In such cases, one must identify the inseparable amalgam of system-instrument.

We should also observe upon this. The more complex the mechanics of the measuring the device – the more complex the M^I that transcribes the $[u', y']$ into $[u, y]$ – the less unquestioning we should be of claims to 'let the data speak for themselves', or, expressed rather more precisely, to 'let the data $[u, y]$ speak the truth in and of themselves'.

Measuring Model Parameters

In the continuum of parameters (*α*) populating a model there are those that are the 'constants in the Laws of Physics', those that can be 'measured independently of the model (*M*)', and all the others, whose values are to be adjusted (or not) in the process of model calibration.

But can some of the parameters be measured *independently*? For strictly speaking, the parameters are essentially the coefficients in relationships between the other more important quantities, i.e. the variables (such as those labelled *u* and *y*) capable of independent observation in the world about us. The parameters have meaning and existence only in *assumed* (or hypothesized) relationships connecting these variables one to another. In other words, and inadvertently (perhaps), some of the relationships in *M* may be assumed to be true in order to back-calculate (estimate) some of the *α* from the supposed 'independent' measurement exercises.

For example, a BOD decay-rate *constant* (*α*) can be back-calculated from the (five-day) BOD bottle test, *assuming* that this process of decay occurs according to the model of what are called linear, first-order chemical reaction kinetics. In effect, and written more generally, something from the model *M* of the environmental system is assumed to be true, in the composition of the model $M^I(α)$ for the 'parameter measurement' exercise, in order to estimate *α* *independently* of *M*, for subsequent substitution of an estimate for *α* back into *M*, to ease and somehow independently 'constrain' calibration of the other parts of *M* (using the field observations [*u,y*]). In truth, the parameter has not been 'measured' in any exercise independent of any scientific hypothesis about the way the world works, including the self-same hypothesis incorporated into the model (*M*) that is to be calibrated.

Propagation of Uncertainty in Problem Solving

First, let us consider 'Given [*u, M*], find *y*!'.

This is forecasting and foresight generation, but only in part. For we shall eventually need to come to a more subtle and more complete view of what constitutes foresight generation (later on pp. 126–33). In this, there might be deep uncertainty not only about *M*, but also about the *future u*, the courses (or scenarios) of future input disturbances, causative stimuli and (external) drivers of the system's behaviour. These scenarios could either be derived from the presumptions, guesses, conjectures, or speculations of people, policymakers, scientists, model-builders, or any others holding a stake in the future behaviour of the system. Alternatively, the future *u* might be generated by another model, of that portion of the world producing *u*, which then recursively returns considerations to the deep uncertainties residing in that other model. It takes little reflection to apprehend the fact that the input to one system (better, sub-system) is always the output from some other sub-system. Any deep uncertainty in *u*, intertwined with that in *M* in the computational mechanics of forecasting, will find its way into the outcomes of those things that are to be forecast, namely the *y*.

We once thought of the weather as such a primal (and unpredictable) driver (*u*) of environmental systems. Its future course would be much as its observed past, albeit shot through with large doses of randomness, the rudimentary cover for our inadequate understanding of the weather sub-system. How times have changed.

Second, there is the problem of 'Given $[M, y]$, find u!', which is variously that of policy formation, management, control and decision-making: what should we choose for those distinctive elements of u that will become the actions, the input stimuli applied *deliberately* to the system to bring about some desired outcome y? Here now deep uncertainty may originate not merely in the science base of M, but in the plural aspirations for the desired outcomes (y), just as we should fully expect, in an argumentative, democratic society. This kind of deep uncertainty, bound to the expression of these plural y, may assume a quite curious, seemingly paradoxical, but unsurprisingly intractable form, as we shall see on pp. 108–17(in more detail).

Given our perhaps scholarly focus on handling (deep) uncertainty in models at the science–policy–society interfaces, gathered, nevertheless, around the practically far more gripping focus of determining elements of u as actionable policies with an impact on the lives of real people, we may note the following in passing. In classical systems analysis, 'Given $[M, y]$, find u!' is often interpreted and computationally articulated as the task of answering the question 'How best?', even under uncertainty. A universe of candidate policy elements of u are to be screened, with the intent to reveal a handful of the more promising, which are then to be evaluated in more detail – in the spirit of 'What if?' conjectures – within the computational mechanics of 'Given $[u, M]$, find y!', with subsequent back-and-forth iterations between the two genres of problem solving. The one, 'Given $[u, M]$, find y!', is a matter of forecasting and forward analysis, the other ('Given $[M, y]$, find u!') is one of back-casting and inverse analysis.[28] This latter has about it a sense of having reviewed what is wanted in the future in order to determine what is to be done now. We shall come to appreciate (on pp. 126–33), how this latter can be interpreted as analysis in support of fashioning 'socially robust' decisions, and not least because of the kind of problem-solving computational scheme (a regionalized sensitivity analysis) designed to work under gross uncertainty about the nature of a system's behaviour.[29]

Change over Time

Just as everything is uncertain, so all things change with time, t, be this a matter of seconds, minutes, hours, days, weeks, years, decades and so on. In principle, all the attributes of modelling are functions of time, so that we have $u(t)$, $M(t)$, $\alpha(t)$, and $y(t)$, where now indeed $M(t)$ entails the {presumed known (t)} and {acknowledged unknown (t)} changing with t.

In the grand sweep of scientific enquiry over the years and decades, in particular, we associate such change with the growth of knowledge. More prosaically, and in the here and now, discerning something about the nature of $M(t)$, for the purpose of generating foresight about future behaviour, has been referred to as forecasting in the presence of structural change ($M(t)$) in the model.[30]

More prosaically too, the fingerprint of uncertainty attaching to the model M prior to system identification – 'Given $[u, y]$, find M!' – will differ from that after the event. Life indeed is a matter of learning, of the endless cycle of ... identification (conjecture) – prediction (refutation) – identification (revised conjecture) – prediction (novel refutation) ... and so on: with some further back and forth in problem solving, now between 'Given $[u, y]$, find M!' and 'Given $[u, M]$, find y!'.

And the Truth of the Matter?

What indeed should be said of the 'truth' of the matter, since a model, by definition, can only but be an approximation of this 'truth'? What does our algebraic language allow us to write succinctly of this?

First, of course, there is the uncertainty that *is* the gap between M and this 'truth' of the matter 'T'. It may seem similar to, if not the same as, MSEU. Yet we have no access to 'T', only uncertain, error-corrupted, imperfect observations $[u, y]$ of the manifestations of 'T', which nevertheless are the well-spring of any attempts to characterize and quantify MSEU. We cannot access 'T' in the absence of uncertainty and error- and bias-corrupted observation, hence the ' ' enclosing the T.[31]

Second, there is the problem of model validation. For the process of solving this problem – of coming to a judgement on the trustworthiness (or not) of the model as a tool for generating foresight on which to found policy – this label of 'validation' has been changed to that of 'evaluation' (but see Box 1).[32] Simply put, 'validation' rather presumed the truth of the scientific hypotheses from the outset. It placed the burden of proof on demonstrating this to be just plain wrong. Application of the associated analytical procedures would err on the side of accepting the model, often as conceived of in the first place. There was a need for a more even-handed approach, not least because the word 'validation' would inspire in the public's mind the rightness and safety of the model. Inevitably, the argument in favour of the word 'evaluation', i.e. the argument against validation, could readily be muddled with Popper's use of *in*validation, itself the basis of the original thinking for the error-driven (but different) process of model structure identification.[33]

The longest chapter in the NRC book on *Models in Environmental Regulatory Decision Making* is that on Evaluation; and it is a subject to which we shall return in greater detail on pp. 126–33.

From the Perspective of Policy

The TransAtlantic Uncertainty Colloquium (TAUC) was funded by the US National Science Foundation (NSF) and US Environmental Protection Agency (EPA). The colloquium ran from 2005 through 2009. Its purpose, defining for this chapter, was to focus *not* on handling error and/or uncertainty in science in *general* at the science–policy interface, but on how to handle the uncertainty arising from the use of complex computational *models* at this interface and, indeed, likewise at the interfaces between science and society and policy and society.

As if timely motivation for TAUC, it was with the following rhetoric that a former Minister of the Environment for Denmark opened the (2004) Symposium of the International Water Association (IWA) on 'Uncertainty and Precaution in Environmental Management' (UPEM): 'What is it about you Anglo Saxons, that you just don't "get it" when it comes to understanding the ideas of the Precautionary Principle?'[34] He was, we may assume, referring to at least UK and US nationals, if not Australian citizens and others. His frustration – we may again speculate – was probably born of the historical differences of the 1970s and 1980s between UK and European Union (EU) attitudes towards water pollution in particular:[35] that the EU saw uncertainty about the behaviour of the environment as cause for man to 'do no harm', whereas the UK was caricatured as a state that would carry on discharging pollutants until the environment audibly and palpably shouted back 'Ouch!'. After all, some of us were schooled in the principle that 'the solution to pollution is dilution', much as the Tees Barrage was originally predicted to achieve.[36]

The original stimulus for TAUC had, in fact, already emerged during preparations for UPEM. This had much to do with the recognition that the US EPA and the Netherlands Environmental Assessment Agency (MNP; now PBL) had for some two decades previously been leaders in the field of handling uncertainty in models at the science–policy interface, yet did not appear to have benefitted greatly from each other's experience. Begun in autumn 2005, TAUC set out to be interdisciplinary by design, with a broad remit, and less focused on the technicalities of computing and assessing model uncertainties than on their consequences in the social processes of policy formation.

Transience or Permanence of Uncertainty

One of TAUC's principal contributions has been to bring legal scholarship to bear on this core problem of handling uncertainty in models at the science–policy interface.[37] The results of such enquiry are summarized in Table 5.1. They were originally drawn up in 2006 by Fisher, Pascual and Wagner. They now appear, further evolved, in Wagner et al. 'Misunderstanding Models in Environmental and Public Health Regulation', where they are being used to dismiss the social construction of models as 'truth-generating machines' or (in their terms) 'answer machines' – for them (Fisher, Pascual, Wagner, that is), strictly in the setting of legal proceedings. Legal scrutiny/review of models in environmental policy in the USA has adhered to the sound science analysis (SSA) paradigm (so labelled in Table 5.1). That in Europe is more indicative of the deliberative problem solving (DPS) paradigm, although the use of models there in developing regulations and policy has hitherto been much more limited than in the USA. As Table 5.1 shows, we note that SSA and DPS would not now be the labels Wagner et al. would attach to the two approaches, for their purposes, to distinguish respectively between rational-instrumental and deliberative-constitutive approaches to public administration.

Table 5.1 reflects a separate line of adaptation (by the present author) since the original 2006 tabulation, albeit now further adjusted in the light of Wagner et al. Our purpose in this chapter is not confined to just the treatment of uncertainty in models in a court of law, as it were, for this would be but a part of the 'complex' at the interfaces among science, policy and society.

Table 5.1: Contrasting features of two alternative schools of thought on how models should be used and judged in the policy process

	Sound science analysis	**Deliberative problem solving**
Legal stance	Rational-instrumental approach to public administration	Deliberative-constitutive approach to public administration
	'Models misunderstood as 'answer' machines [truth-generating machines]'[38]	'Models properly understood'[39]
Uncertainty	Undesirable, but transient feature	Inherent, enduring feature
Expectation of model	Representation (ultimately) of reality	Tool, analogy or metaphor
Purpose	Prove regulation is supported by 'sound science'	Eliminate daft ideas sooner than later, especially in a multi-disciplinary (if not trans-disciplinary) context
Accountability	Proximity of model to reality	Effective policy-problem solving
Expectation of modellers and analysts	Delivery of detached, objective analysis	'Clinical' neutral detachment from the problem is illusory (according to this author's experience); subjective choices are made
Stakeholders	Disputation/undermining of government science	Facilitate deliberation, constructive contestation, and learning (if not negotiation)

The second row of Table 5.1 distinguishes the two paradigms, or schools of thought, with respect to 'uncertainty'. Even a reader with little more than just casual contact with modelling over the past four decades, or with the majority of professional environmental modellers, will know well enough this assertion: with a larger computer, the uncertainties in models will be reduced. The aspiration endures, for example, in this summary of a 2010 meeting at the (UK) Royal Society on 'Handling Uncertainty in Science':

> [H]ow do we value the loss of the Amazonian rainforest, or of large parts of Bangladesh, or of prolonged Sahalian drought?
> [...]
> Estimating value in this generalized sense, including the thorny (and ultimately ethical) issue of whether the suffering of future generations should be somehow discounted, is clearly an extremely challenging issue for all of us. Nevertheless, these challenges should not deflect scientists and governments alike from ensuring that we are doing all that is humanly possible; firstly, to estimate uncertainties in future climate change as accurately as possible, and secondly to reduce these uncertainties – a large element of which lies in improving the computational representations of the equations of climate – wherever we can.[40]

If a larger computer enabled the construction of a larger model, we might conjecture, and one in which all constants (α) were solely those of the laws of physics – thus doing utterly away with model calibration (and model structure identification, for that matter) – this would promise a very grand culmination, a happy state of affairs with truth-generating and answer machines (not, of course, of the old telephonic variety).

Defining in Table 5.1 is therefore the guiding principle of the SSA school of thought: that uncertainty is transient, reducible to insignificance. In the limit, we might say: 'The uncertainty in the {acknowledged unknown (t)} tends to 0.0 as t tends to infinity (∞), such that $M(t)$ tends likewise to 'T' as $t \to \infty$'. Such an outcome is precisely the 'expectation of the model' under the SSA paradigm (Table 5.1). Over time, the {acknowledged unknown} will have been transferred into a flawless {presumed known}. And all those engaged in the construction and use of the model for the 'purpose' of forming regulatory policy will be held to 'account' on the kernel property of the trustworthiness of the model, i.e. that the uncertainty regarding how well M approximates 'T' is below some threshold of acceptability.

This – the expectation of {acknowledged unknown (t)} $\to 0.0$ as $t \to \infty$ – is not the view expressed in the recently completed *White Paper* for the US National Science Foundation (NSF) on *Grand Challenges of the Future for Environmental Modeling*.[41] We are some considerable distance from having a 'predictive science of the biosphere' – since the question of whether we have this has, in fact, been asked[42] – and philosophically perhaps destined ever to be so, more in line with the DPS paradigm, of uncertainty as everlasting.

Agreement: The Continuum from Consensus to Conflict

Unsurprisingly, there can be disagreement, including (presumably) on whether working within the SSA or DPS paradigms of Table 5.1 is the 'right' way to handle uncertainty. For example, commenting on procedures within the Intergovernmental Panel on Climate Change (IPCC), Patt[43] has observed how that process takes better account of the mathematical and statistical kinds of uncertainty than the uncertainties arising from disagreements among experts, i.e. professional scientists, as echoed in Smith and Petersen (this volume).

Writing earlier on *Uncertainty and Quality in Science for Policy*, Funtowicz and Ravetz[44] introduced what they called a 'research-pedigree matrix'. As a field of enquiry matures, they argued, reliability in the status of the relevant science evolves through the following stages:

> from 'no opinion' with no peer acceptance;
> through an 'embryonic field' attracting low acceptance by peers;
> 'competing schools', with medium acceptance;
> a 'theoretically-based model' accepted by 'all but rebels'; and on, in the end, to
> an 'established theory' accepted by 'all but cranks'.

The expressions quoted are the words of Funtowicz and Ravetz. The inference in this progression is that competing schools will – or should – eventually yield to the orthodoxy of a single school of thought.

In the beginning, no opinion (O) is held by any scientist first turning her or his attention to the problem at hand. Nor has much, if any, evidence (E) on the nature of the problem been acquired in a scientific manner. There are few scientific facts as such. As the status of competing schools is attained, there is evidence (E) and more than one roughly equally tenable opinion (O) on the subject matter, if not a multiplicity of them. Dissonance and disagreement would appear to be at their highest under this state of affairs. Having risen thus to a peak, one may suppose that dissonance slowly subsides thereafter, as the state of competing schools is harmonized, to colleague consensus, which consensus itself then hardens into the certainties of rejecting the rebels, and eventually into dismissing as quite irrational any remaining cranks. Kuhn's Normal Science may thus proceed smoothly on its way.[45]

Over time, we can see that the nature of the opinions has evolved: from originally no opinion about the nature of the problem to be solved; to, in the end, and in the first and dominating rank, an orthodox opinion on both the nature of the problem and the means of its solution; *and*, in the inferior second and third ranks, to the opinions of the rebels and cranks, respectively, which are increasingly ridiculed by the orthodox group of consensual colleagues, who consider each other the only scientific peers, essentially members of just their own group.

This is a gross (but insightful) simplification and generalization, no doubt. We may imagine it applies to each and every single discipline within science.

What, we may ask, therefore, might obtain when evidence (E) and opinion (O) relating to several disciplines must be taken into account and when the nature of the problems are of substantial and immediate public interest? For these are the prevailing conditions of forming policy for environmental stewardship.

In December 2003, Sheila Jasanoff responded formally to a request for comments on peer review protocols issued by the Office of Management and Budget (OMB) of the White House (her comments are posted at www.whitehouse.gov/sites/default/files/omb/inforeg). The request was phrased as follows:

> On August 29, 2003, OMB's Office of Information and Regulatory Affairs (OIRA) issued a *Proposed Bulletin on Peer Review and Information Quality*. The purpose of the *Bulletin* is to ensure 'meaningful peer review' of science pertaining to regulation, as part of an 'ongoing effort to improve the quality, objectivity, utility, and integrity of information disseminated by the federal government'.

Dividing science into 'research science' (as in the foregoing) and 'regulatory science' (an important part of the concerns of this chapter), and drawing upon both her own work (in *The Fifth Branch*) and that of Funtowicz and Ravetz,[46] Jasanoff begins her response as follows:

> When science is 'normal' or paradigmatic in the sense described by the philosopher of science Thomas Kuhn, independent review can help ensure that researchers are applying the standards of their field rigorously, consistently, and without bias or deception. In these circumstances, there is ordinarily little doubt who counts as a *peer*. Peers are the recognized members of the scientific specialty or subspecialty within which normal science is conducted. Such peers share a common culture of scientific practice, with a shared commitment to the goals and methods of inquiry in their field.

Looking back to Table 5.1, along the row beginning 'Expectation of modellers', such sentiments resonate with the SSA paradigm, with its 'detached, objective analysis'. Jasanoff goes on to say:

> Regulatory science, however, is not normal science. It may cross disciplinary lines, enter into previously unknown investigative territories, and require the deployment of new methods, instruments, protocols, and experimental systems. Correspondingly, the 'peers' for reviewing regulatory science are likely to come from disparate technical backgrounds and not form part of a single, tightly-knit research community.

This, now, lines up much more with the DPS programme of Table 5.1. Personal experience, including a threat of legal action, has made it abundantly apparent that the detached, academic analyst can become a part of the problem, hence the import of the entry – regarding this author's personal experience – along this line ('Expectation of modellers') in Table 5.1.[47]

And so we come to Jasanoff's punchline: 'Many academic observers have suggested that regulatory science in areas of high uncertainty should be subject to wider and more public critique – sometimes termed "extended peer review" – rather than to traditional peer review by technical experts alone'. The complex, computational models M of environmental science are inherently multi-disciplinary. And their extended peer community clearly spills over and out of science to encompass policymakers, industry/business associations (as affected directly by the proposed regulations), members of non-governmental, environmental activist groups, and members of the general public. For Jasanoff, her comments dealt with uncertainty in science in general at the science–policy interface. For the Netherlands Environmental Assessment Agency (PBL), the issues of handling uncertainty in *models* at the science–policy–*society* interfaces have become the reality of its everyday practice.[48]

Multiplicity and Variety of Stakeholders

To extend the language of our algebra, there can in principle be a proliferation of opinions, denoted O_i, as held by stakeholder community i, where $i = 1, 2 \dots m$, and where m is possibly quite a large number.

Competition, disagreement and the absence of consensus are probably the norm. Indeed, if cultivated and harnessed among opinions O_i that are granted roughly equal validity in the debating chambers of policy formation, competition as such promises the great good of 'high deliberative quality' in what Ney[49] presents as a refurbishment of Dahl's pluralist democracy.[50] The status of fairly competing schools of thought seems most attractive under this view. The difficulty – including within the IPCC process of coming to terms with climate change – is that all too often the voices of the competing schools of thought are not accorded equal respect, much as scientific orthodoxy will work to silence the rebels and cranks. Here, for example, is Patrick Michaels of the Cato Institute commenting upon the conduct of the IPCC's Chair (posted at www.forbes.com on 21 April 2011; last accessed 10 December 2012):

> In 2007, the UN famously stated that, if warming continued at present rates ... the massive Himalayan glaciers would disappear 23 years from now. While the source, the UN's Intergovernmental Panel on Climate Change (IPCC) proclaims itself the consensus of climate science, there's no credentialed climatologist on earth who believes that this ice cap, which is hundreds of feet thick, could possibly disappear so soon.
> [...]
> When the government of India, which knows something about the Himalayan glaciers that feed the great Ganges River, challenged the UN's forecast, the head of the IPCC, Rajenda Pachauri, labeled it 'voodoo science'.[51]

For things to mature through the pedigree matrix of Funtowicz and Ravetz – from competing schools to the consolidated, single orthodoxy of colleague consensus – the resulting overbearing, hegemonic (bold-faced) **O** may not be entirely what we should wish for. Better perhaps, some state of competing schools of thought should prevail. In the spirit of *Cultural Theory*, whose seminal text was also published in 1990,[52] if there is an orthodoxy, then even the voices of the rebels and cranks should not be entirely ignored, although their opinions might be diminished in stature (small 'o' and even smaller $_\circ$, as it were, in the eyes of the orthodoxy, **O**). After all, whence derives the anomaly and its irritant advocate in research science – the rebel or crank – who motivates the paradigm shift,[53] to set the cycle off again? Uncertainty, and much of it, may return to re-assert itself over the given field in which the revolution is taking place, with the hardened opinions of the 'masters and mistresses' of the former predominant consensus school seeming themselves now ever more anachronistic, if not crank-like.

All this argumentation, however, is taking place in just the domain of Jasanoff's research science.[54] In regulatory science, the government determines and promulgates the regulations. And as Table 5.1 shows, under the SSA paradigm, it is the intent of some of the 'stakeholders' (outside of government) to dispute and undermine the 'government science', which we can gainfully label as model M_G, where subscript G denotes government. Typically, those to be regulated by EPA rulings, such as industries and businesses, will claim that *their* science, M_R (with subscript R for the 'regulated'), involves nothing like the future implied by M_G. What is more, *their* science is superior to the government's, not least in its being far less uncertain, if not indeed passed off by the R community as certain.

Uncertainty: Deepening by Degrees

We need yet more algebra, to define more crisply the archetypal theory of making a decision: to build, or not to build, the Tees Barrage; or there again – in an exemplary study of making decisions under uncertainty – to build, or not to build, a barrage across the Oosterschelde estuary in the Netherlands.[55] Roughly and briefly stated, in a deterministic world (in fact, that of a deterministic model M), the commercial mussel industry of the inner Oosterschelde could have expected building of the barrage to bring a bigger catch. In an uncertain world, that bigger catch was forecast to come with a not-insignificant risk of the complete collapse of the mussel population, which would otherwise have been entirely absent from the foresight of the deterministic analysis.[56]

A decision is taken to be an action applied to the system, as an input (u), to influence that system's behaviour in bringing about some desired output response, now bearing the more precisely qualified label of y_d. To distinguish this decision from all the other factors, i.e. the incoming disturbances buffeting the behaviour of the system back and forth, we use the label u^*, with the asterisk denoting the not uncommon aspiration to achieve some kind of 'optimality' about the decision (as the answer to 'How best?'). As we stand on the threshold of the future, let us encapsulate in the abstraction of a 'random event', r, everything that could cause pollution in truth to be out to sea or against the

Tees barrage, or cause the mussel population to proliferate or collapse within the Oosterschelde estuary. In short, in general, and in the abstract, this is to say: encapsulated in the archetypal and notational simplification of r, is everything that could cause the decision-event couple $(u^*\text{-}r)$ to have n outcomes, each with its respective probability of occurrence, p_j, where $j = 1, 2 \dots n$. In a deterministic world, for contrast, r technically has but one outcome, with probability 1.0.

Formal, mathematical analysis for identifying u^* is generally referred to as decision making under uncertainty (DMUU).[57] Within this idealized framework, uncertainty surrounding the analysis of a decision can be classified into three significantly different categories, deepening by degrees, step by step:[58]

(U1) The exhaustive set of discrete (n) possible outcomes of the event (the future states of nature) is known, as too are the probabilities of occurrence (p_j) of each outcome; this has been referred to as Statistical Uncertainty.

(U2) The exhaustive set of outcomes is known, but not all of the outcome probabilities, i.e., Scenario Uncertainty. Now there is uncertainty about the uncertainty, i.e., about the nature of the set of p_j, which (if one thinks about it) is the beginning of an infinite regress.[59]

(U3) Not all of the outcomes are known, *ergo* the set of probabilities cannot be known, i.e. a state of what is called Ignorance prevails.

Yet even the category of Ignorance (U3) can be sub-divided along the lines of increasing uncertainty,[60] expressed colloquially, albeit not in the Rumsfeldian style,[61] as: 'We don't know what we do not know'; 'We will never know'; to 'We cannot know'. From some point along this continuum of deepening degrees of uncertainty, we can choose to single out a fourth in our categories of uncertainty, as follows. With multiple stakeholders disputing each other's versions of the {presumed knowns}, and each pointing to all that they discern as the {acknowledged unknowns} of the others' science, so great is the uncertainty in the decision framing that:

(U4) More than one version of the truth of the matter is actively maintained and promoted, by each of those who hold a stake in the outcome of the decision-event couple $(u^*\text{-}r)$. Each such alternative, however, is caricatured as having the *certainty* of but a single outcome. Given the differing convictions of each group of stakeholders about the way the world works, no longer is there any remaining shred of the uncertainty of 'everything that could cause pollution in truth to be out to sea or against the Tees barrage' (or at the $(n - 2)$ points in between, for that matter). It will be in just *one* place, with probability 1.0, i.e., with certainty. The state of affairs is that of a *plurality* of Contradictory Certainties,[62] i.e., decision making under contradictory certainties (DMUCC).

Here (with some exaggeration) we are in a situation of arch disagreement: 'What I know is the truth; what you know is utterly false'. This is readily recognizable as a euphemism for disagreement, perhaps not so much about the model and science from which it is drawn, but about what is desired as the outcome of the decision context, born of profoundly differing views (for us in this chapter) on the man–environment relationship. Were a problem to lie within this seemingly paradoxical situation of contradictory certainties, we might find a plurality of such statements, each buttressed indeed by a quite different model M_p, but a 'certain' M_p in the eyes of its proponent – one for each stakeholder community i.

An elementary illustration of this has already surfaced, in the plural rationalities of: first, the government agency that fashions the (provisional) environmental policy (u^*), based on a model, which then becomes labelled, if not denigrated (the closer the hint of a legal discourse), as the 'government model', M_G; and, second, the regulated group (such as an industry association), who, perceiving themselves to be threatened by the provisional u^*, promote *their* model M_R as the more trustworthy mechanization of the scientific knowledge base.

To generalize, the nearer draws any legally framed discourse for resolving the disagreement – indeed dispute – the less uncertainty any solidarity (i) of decision-stakeholders will attach to its construction of the truth, i.e., its convictions about the way the world works. In effect, in the abstract, the narrower and sharper becomes the probability distribution assigned to the outcomes of the decision-event couple (u^*-r). In the limit, each M_p for the $i = 1, 2 \ldots m$ groups, is held by each respective i to be certain, while, from the stance of that group i, all the other contending accounts of the situation – all the other (m-1) models M_q, where q does not equal i – are becoming more unacceptably uncertain with every step towards the metaphorical or literal court of law. The fiercer too becomes the competition under the otherwise desirable state of Funtowicz and Ravetz's competing schools of thought.[63] We might even observe that their 'rebels' and 'cranks' are not irrational, as claimed from the perspective of the orthodoxy. Members of these belittled ranks simply reason (coherently, perhaps cogently) about the world in ways quite different from those of the orthodoxy – in other words, we have the circumstances of plural rationalities, i.e. cultural theory.[64] *Dissentio, ergo sum* ('I disagree, therefore I am'), is how adherents of this cultural theory would restyle the well-known aphorism defining our existence.[65]

Perhaps this state of affairs (U4) should not be subsumed under the rubric of uncertainty. Perhaps we should not have presented it as the sharpest of plural certainties constituting the height of uncertainty, with yet its inherent paradoxical nature. But it is an all-too-familiar situation. DMUCC is significant because, first, it may better approximate the circumstances of many decision-making problems in policy formation and reform and,[66] second, because it has attracted little analysis in respect of the employment of typical, complex environmental models (M), with the rare exception of van Asselt and Rotmans.[67] In any case, DMUCC has been embedded as the nub of another of the grand challenges (no. 10, in fact) of Beck et al. (2009), hence rightly the complement of the challenge (no. 7) of model structure identification wherever there is deep uncertainty in the science bases (from the second section, pp. 101–3).

Taking Stock

We began with the separation of the {presumed known (*t*)} and the {acknowledged unknown (*t*)}, and the fluidity and interchange of their contents over time *t*. We used these notions to distinguish between two strategically different ways of viewing uncertainty in models at the science–policy interface, namely those of sound science analysis and deliberative problem solving (in Table 5.1). Within the pedigree matrix of Funtowicz and Ravetz,[68] we suggested the state of competing schools of thought should be privileged. This may be frustrating, inelegant, and not a state of consensus. Yet the supposedly self-evident virtue of consensus itself has in any case been vigorously disputed, especially from a legal perspective.[69] This again is the way of the world.

Taking once more the long view, over the sweep of scientific history, we have drawn upon the work of Jasanoff's *The Fifth Branch* to recognize the distinction that has emerged between how quality is maintained in research science and regulatory science, to bring attention to the differences in peer review – symbolized as the difference between an opinion (**O**) from the scientific orthodoxy and that of the scientific crank (with his/her most diminutive $_{o}$, in the eyes, that is, of the over-bearing majority **O**). The telling point of Jasanoff's argument is the idea of extended peer review for our regulatory science herein,[70] with potentially its welter of opinions (O, *i* = 1, 2 ... *m*), from just about all quarters of society and the policy-forming communities, beyond the confines of science. As Funtowicz and Ravetz would have it,[71] along with a growing number of others,[72] we are in a state of post-normal science (PNS), where the 'facts are uncertain, values in dispute, stakes high and decisions urgent'.

To conclude, we have mapped some of the contours of the anatomy of this problem of handling uncertainty in *models* at the science–policy interface. And we shall return below (on pp. 126–33) to the matters of how we might approach solutions to some of these problems, in the setting of DMUCC, in the deepest depths of uncertainty: in matters of generating foresight, about whether a policy is needed in the first place; in coming to a judgement on the trustworthiness of a model on which to found policy; and, subsequently, in adapting the intent of laws, policies, and decisions already put into action.[73]

From the Perspective of Society

Half a century ago, at the turn of the 1950s/1960s, there was no precautionary principle for we Anglo Saxons to 'get', at least not those of us in the UK (references to the principle began to appear in the late 1970s).[74] Science was laboratory or research science, not regulatory science. It was normal, albeit occasionally revolutionary, but not yet post-normal. Peer review of the works of scientists was conducted strictly by other scientists, certainly not by scientifically lay members of society beyond the confines of science.

The discipline of environmental science, though immature, was nevertheless evolving along a short path to becoming replete with ever more complex, high-order computational models M. Two texts on water pollution control, which by the turn of the 1970s/1980s could reasonably claim sufficient perspective to capture the mood of the preceding decade or so,[75] confirm essentially the complete absence of the human dimension from the scope of any computational M of the 1970s. Even in those models designed for the purpose of 'Given $[M, y]$, find u!', where one might most have expected it, the human dimension did not gain access. No mathematically expressed utility function, which would imply some attitude towards the risks (*ergo* a human dimension) associated with attainment (or otherwise) of the desired outcomes (y) by the decision (u), is apparent in any of the books' computational models (M).

By the late 1990s, simulated human agents imbued with the plural rationalities of *Cultural Theory*[76] – with its typology of plural attitudes towards risks among the various social solidarities to which those agents might belong – had become embedded with facility in computational models of farmer-landscape-lake systems.[77] With the advent of extended peer review, scientifically lay members of society (beyond the boundaries of science) were invited to exercise their powers of review over scientific models. These same ordinary people could come to find their behaviour simulated in those same scientific models. No wonder they should now hold a stake in the trustworthiness of such a model!

As with the human dimension, so too was 'climate' absent from the subject indexes (hence the models) of these early books on environmental modelling.[78] Pollution of the rivers of Europe and North America was a palpable fact long before there were computational models – not so climate change. For this was how the US Environmental Protection Agency (EPA) saw the need (a decade or so later) for the development of an earth systems model, unmistakably an M:

> The global climate change problem is essentially one of the expectation of climate change in response to increasing atmospheric greenhouse gases. Were it not for these expectations, which have always been derived from theoretical considerations, i.e., some form of model, we would perceive no problem.[79]

The difference is as that between what we customarily call an indisputable historical fact, e.g. water pollution, and what most of us perceive as an oncoming threat, namely, climate change and greater climate variability.

Things are in a state of flux; and ever was it thus. They are also in something of a state of mounting technical complexity and greater social turmoil than once was the case in shaping environmental policy. Drawing upon van der Sluijs's account[80] of the outcomes of the 2004 UPEM Symposium, Curry and Webster write of 'Climate Science and the Uncertainty Monster'. The term (monster) was borrowed by van der Sluijs from the Dutch philosopher Smits, who in turn had built upon the seminal work of anthropologist Mary Douglas (from the 1960s), itself the cornerstone of cultural theory.[81] Their papers (Curry and Webster; van der Sluijs) paint a vivid picture of the labyrinth of social and political games that are played in the handling of uncertainty in science – in general – at the science–policy–society interfaces.[82] Our purpose herein, however, is not that of mapping that particular labyrinth, hence finding a way through it to the actionable decision.[83] Rather it is to reflect (for the time being) on these lesser questions:

> Consider the (simplified) nature of environmental policy formation half a century ago, consider it today, and juxtapose the two. How has the projection of both models (M) and modellers into the 'complex' at the interfaces among science, policy, and society changed there the nature of problem solving and decision making?
> [...]
> More specifically, what quintessentially is the difference – if any – between handling uncertainty in science in general and handling deep uncertainty in models, in particular, at these interfaces?

The Technical Material of Evidence-Based Policy

Once was the time – fifty years ago, we submit – when the formation of environmental policy was founded upon just two caricatured (and doubtless) simplified categories of information:

1960s: (E1) Evidence from past scientific laboratory experiments and scientific field monitoring, i.e., past $[u, y; t^-]$, where t^- signals 'time past', coupled with (O1) scientific opinion about implications for the future, i.e., about future $[u, y; t^+]$, in which t^+ signals 'time future' – but *not* any facet of the future generated from a computational model, as we would know it today.

Scientific opinion under category (O1), we further submit, would have been akin to the 'theoretical considerations' alluded to above in Mulkey's reasoning about how we might (previously) have arrived at the 'expectation of climate change'.[84] In turn, what he had in mind for these theoretical considerations – 'some form of model' – is redolent of the kind of model used by economists to reason with, *without* a computer (or, strictly speaking, without an M), as opposed to reasoning with words, as we observed in introducing this chapter.[85]

To these two time-honoured categories of evidence and opinion must now be added:

2010s: (E2) Evidence of how the model (M) mimics the past behaviour ($[u, y; t^-]$) of the real thing, i.e., evidence of the match (or otherwise) of M *versus* 'T', or $[M \ vs \ 'T']$ for short, and (E3) scientific evidence about the future $[u, y; t^+]$ generated from M (as discussed on pp. 101–7), coupled with (O2) scientific opinion about the trustworthiness of the model (M) as a predictive device and (O3) extended peer opinion (in the style of Jasanoff and Funtowicz and Ravetz), likewise, about the trustworthiness of the model (M) as a predictive device.

Here, by design, the algebraic symbol for the model (M) has been displayed at every opportunity, to emphasize the difference that is of core importance.

The contours of this difference can be plotted along three dimensions:

(D1)   Each 'agent', or decision-stakeholder, can (and will) hold an opinion on each category of evidence (E) and, of course, on the opinions (O) of every other stakeholder, of whom there are now many more than fifty years ago.

(D2)   Merely as a function of the advent of computational models (M), stakeholders from the scientific community can now enjoy the benefits of sub-division and differentiation: first, between those who develop and use the M, and those who (often avowedly and obdurately) do not; and, second, within the former, between those who believe in the validity only of 'physics-based', 'large' models (M) and those who do not, for they value (above all) 'small' models, i.e., essentially 'statistical' models, in common parlance.[86]

(D3)   Merely as a function too of the advent of extended peer review, an entirely new set of stakeholders has been invited into the process of coming to a decision, to the shaping of actionable policy.

Not necessarily orthogonal to (D2), and perhaps not a dimension in its own right, we can today recognize the separation of postures, if not identities, into the threesome of the scientifically orthodox solidarity, the scientific rebels, and the scientific cranks. But these distinctions were always there in science.[87] They are accordingly not a modern phenomenon or – significantly for us – a differentiation induced by the introduction of computational models into environmental science over the past half century.

Our affairs have become technically more complex, then. We say 'technically', because (again, in the interests of simplicity) we wish to distinguish between the blocks/categories of information and insights (E and O) germane to the making of the decision and the socio-political games that will be played by the decision-stakeholders in the process of deploying and manipulating this information in coming to a decision. Identifying, and even creating and managing the rules of this gaming, is well beyond the scope of this chapter, although Wagner, Fisher and Pascual,[88] to their great credit, have begun to engage with this challenge (within the domain of legal scrutiny and debate of models). What is more, having now rather sharply separated the evidence (E) from opinion (O), one cannot help wondering about the role of O in forming *evidence-* or *fact*-based policies, either then (in the 1960s) or now.

Models and Modellers: Insurgents Both

We know how members of the scientifically orthodox solidarity (in just a single discipline) will expect their opinions (**O**) to outweigh any other opinion, in particular, those of the scientific rebels and cranks, on any and all evidence (E) and opinion (O) deemed to be scientific. We have already ranked these opinions symbolically: as **O**, o, and $_o$, respectively – a ranking, in fact, against which the rebel should rebel, and which the crank ought to dismiss as madness. *Dissentio, ergo sum* indeed. We have seen the accompanying exercise of 'power', in the accusation of voodoo science from the chair of the IPCC (in the foregoing section). People are free with their opinions. How telling is the phrase 'intellectual pecking order'. How free we are in taking every opportunity to remind each other of it.

There are places in science, nonetheless, where restraint remains the custom. Yet the decency of such restraint is actually a source of great difficulty, precisely because of the modern-day use of complex, hence necessarily multi-disciplinary, computational models **M**. Scientists coming from one discipline, let us call it k, with their opinions O_k, will rightly hesitate to pontificate on the opinions of scientists coming from another discipline, l say, i.e., O_l. This is indeed technically quite a problem for (scientific) peer review – in the round – of our contemporary, multi-disciplinary models, wherein technical material from discipline k has to be seamlessly and correctly fused with that from discipline l. The fisheries ecologist rightly hesitates to opine on the quality of the bits of the science base the hydrodynamicist has judged appropriate to incorporate into, say, the model of the Oosterschelde estuary. How then shall we know whether all the many multi-disciplinary *parts* of the model **M** add up to something of a trustworthy *whole*, once all the parts have been duly pinned together in the computer program? Where is the polymath, or computer-literate 'renaissance man', in whose opinion on the integral whole of **M** we should place our trust?

Such inter-disciplinary restraint is far from universal, of course. Here is how the Preface to the book *Environmental Foresight and Models* begins:

> Everyone – or so it seems – has an opinion on modelling. This belittles the subject, I fear. For it suggests the principles of constructing a mathematical model, evaluating it against whatever experience one has of the behaviour of the prototype, and subsequently applying it for the purposes of exploring the future, are trivially straightforward – easily grasped by the non-expert.[89]

The defensiveness in this is quite apparent. It was born, in part, of the personal experience of the UK-Scandinavian Surface Waters Acidification Programme (SWAP) of the 1980s.[90] In this, there were the modellers, constructing their **M** in their (comfortable) London offices. And then there were the field scientists, dedicated to the (necessarily) all-weather pursuit of gathering in the data $[u, y]$, from the rugged open spaces of Scotland. For these non-modeller scientists, possession of the data was 'nine-tenths' of the right – with an allusion to the role of 'possession' within the law — of being the first to publish their work in a peer-reviewed scientific journal. Thus could power be exercised by one scientific solidarity over another. It is hard to find the **M** in 'Given $[u, y]$, find **M**!' when one has not, in fact, been given the $[u, y]$!

Modellers *are* resented, not least by other scientists; and perhaps because of the threat they pose to the pre-existing arrangement for the brokering of power in the processes of making public policy. They are the artisans of *Useless Arithmetic* in the eyes of some.[91] The dialectic is sharply expressed as that between 'to trust or not the M'.

Computational models, and the interpretations and forecasts derived from them, have become matters of both very public debate and popular concern. To his 2004 novel *State of Fear*, bestselling author Michael Crichton appends an illuminating 'Author's Message', in which he offers his perspective on the state of play in modelling and forecasting the impact of climate change on sea levels:[92] 'We need more people working in the field, in the actual environment, and fewer people behind computer screens', he urges. Yet even if the level of effort devoted to both sides (observation and computation) were balanced, the divide might not be bridged. Indeed, what transpires at this seemingly esoteric divide – within the nub of the deep (hence potentially obscure) issue of model structure identification (as defined on pp. 101–7; and also as grand challenge no. 7 in Beck et al., 2009) – can turn out to be both very public and highly contentious. Mooney's popular account of *Storm World – Hurricanes, Politics, and the Battle Over Global Warming*, is cast exactly there, with characters to mirror the divide: MIT scientist Kerry Emanuel and colleagues set in the computational camp with their models (M); Colorado State University meteorologist William Gray and associates cast as empiricists, who plead for the data to 'speak for themselves'. It makes for good reading to pit the two camps against each other, in this case without apparent inaccuracy in reporting. There appears to have been no meeting of minds; no reconciliation of the archly opposed positions; no reconciliation (in our 'model world') of M with $[u, y; t^-]$; no constructive resolution of some of the deep uncertainty in M as a result of the confrontation.

Things can get worse. In 1999, pandemonium broke out in the normally quiet world of environmental foresight in the Netherlands. Its National Institute for Public Health and the Environment (RIVM), officially charged with preparing the country's State of the Environment reports, was publicly accused of lies, deceit and shoddy workmanship with its computer models (M) – by one of its own statisticians.[93] The affair became front-page news, received prime-time coverage on television, and provoked questions and debate in the Dutch parliament. As Petersen et al. recall: 'After years of trying to convince his superiors that the agency's environmental assessments leaned too much toward computer simulation at the expense of measurements, [the statistician] went public.'[94] The upshot has been to put PNS into practice,[95] the foundations for which had been laid years before, going back to the 1980s[96] – hence the significance of those very closely associated with this experience participating in TAUC.

Such uncloaking and public exposure of the weaknesses of models is neither new nor about to cease. As if to echo the Dutch (RIVM) 'foresight scandal', models, it has been said,[97] allow the craft skills and expertise of the model-builder to be legitimated – made objective (as opposed to subjective) – such that that expertise may be presented in an impersonal manner. Some would say disparagingly 'passed off as detached and impersonal'. Model-building is thus merely the latest in a long tradition of creating oracles to be consulted. Deliberate shrouding of the soothsayer's device in a mystique is ages-old, Schaffer[98] would remind us, doubtless to the delight of Pilkey and Pilkey-Jarvis, and just as Green and Armstrong would have supposed:[99]

> The forecasts in the [2007 IPCC WG1] Report were not the outcome of scientific procedures. In effect, they were the opinions of scientists [(O)] transformed by mathematics and obscured by complex writing. Research on forecasting has shown that experts' predictions are not useful in situations involving uncertainty and complexity. We have been unable to identify any scientific forecasts of global warming. Claims that the Earth will get warmer have no more credence than saying that it will get colder.

Indeed, as Smith and Petersen (this volume) present this, the 'relevant dominant uncertainty', or 'weakest link' in the chain of activities in coming to the decision at hand, may well lie quite outside the frame of any model. More troubling, it may lie in the scientific evidence either *not* considered in the processes of deliberation or in that even *suppressed* from public declarations by members of the community of climate modellers. Smith and Petersen (this volume) are especially sceptical of the manner and phrases in which climate modellers responsible for the UK Climate Projections of 2009 (UKCP09)[100] report to the wider audiences upon the uncertainties attaching to their computational exercises.

We all belong to a community, perhaps several of them. Each community has its prejudices. And models are not always viewed in a favourable light, as is now abundantly apparent. Since some would be much less detached in expressing their resentment of models and modellers, it is clear that our community suffers from a problem of 'image'. We should be obliged to attempt to overcome it.

Then and Now: The Essential Difference

Technically speaking (once more), and to reiterate, the differences between then (the 1960s) and now (the 2010s) are these:

(T1) The volume of conventional, time-honoured, traditional evidence available (E1), i.e., past data $[u, y; t^-]$, has hugely expanded.

(T2) To this expanded volume of conventional evidence has been added the previously never encountered volumes of (numerical) evidence (E3) on future data $[u, y; t^+]$ and (E2) on model evaluation, i.e., $[M \ vs \ 'T']$. 'Big data', comprising all three, (E1), (E2), and (E3), is today a popular phrase.

(T3)   Given extended peer review and the professional sub-divisions in science induced by the arrival of *M*, the number of differentiated players, all with their opinions ((O1) through (O3)) on the various blocks of scientific evidence, has swelled substantially.

(T4)   In particular, and irrespective of all else that may have changed or stayed the same over this half century, a novel category, breed, or specialized discipline of scientist has arisen: s/he necessarily associated with the rise of *M*, their development, their evaluation, and their application.

On these four accounts, the problem-solving domain of concern in this chapter has at the very least become technically, procedurally and socially more complicated.

But there is more, of course. Some of the vital, if not existential, problems to be solved have changed: from problem solving associated with established, historical facts (of the past) to problem solving associated with conjectured more or less likely oncoming threats (of the future), and not just climate change (witness the various case studies in *Late Lessons From Early Warnings*).[101] We recognize, too, that engaged in such conjecturing is not merely the orthodox scientific solidarity, but also those of the rebels and the cranks. Indeed, that these latter are so conjecturing is to be applauded, since consensus alone is anathema to exploring the richness of our possible futures.[102] If such an 'issue-shift' over the past half century is correct, some of the accompanying burden of significance – in the science and scientists contributing to our committing to a mistake in policymaking – has shifted: away from the exclusive, hence supreme, importance of past data, and away from the exclusively privileged opinion of the laboratory and field scientist; to the increased importance of future data and model evaluation, and to the therefore legitimate opinions of the relevant professional scientific experts – the modellers – on these new blocks of evidence. Because of the relevance of more than just the past data (E1) and the attaching expertise (O1), the standing of this evidence and attaching opinion (of the 1960s) can only but have been diminished over the past five decades, merely by the 'dilution' in what is taken into account (if for no other reason).

Within the realm of opinion (O), and notably *not* opinions on the evidence (E), but on the various other players within the scene, and the expertise (if any) they bring to the table – especially in respect of scientific expert opinion on another scientific expert's opinion – *I* (first-person singular) have been startled at the vehemence of the resentment shown towards modellers at a seminar convened within the scope of TAUC and hosted in Oxford in 2008.[103] It seemed to go beyond what one might have expected from exercise merely of the intellectual pecking order within science. With even an element of drama about them, there were calls for modellers and their models to be 'outed': cast into the harshest of spotlights from the media; and required to justify their existence, in terms more articulate than modellers are customarily credited with commanding, as in a parliamentary or congressional committee hearing.[104] The evident sharpness of this divide between modellers and non-modeller scientists was somewhat unexpected. One can sense much of the same in Smith and Petersen (this volume), where the third of Smith and Petersen's three dimensions of *reliability* has to do with the 'extent to which modellers are trusted'.

And What of Deep Uncertainty?

The deep uncertainty in models (described above on pp. 101–7) will shape the contents of the categories of evidence associated with model evaluation ($[M \ vs \ `T`]$) and future data ($[u, y; t^+]$), especially where the future might be 'radically different' from the past, as opposed to 'essentially similar' to it. To be pedantic about this, given no past evidence ($[u, y; t^-]$) by which to corroborate or refute any parts of M representing the computational mechanics (and science) of how radically different behaviour might be generated, great uncertainty ought in principle to attach to these parts (some of the α) after system identification – after reconciling M, that is, with the intrinsically inadequate $[u, y; t^-]$. The indelible fingerprint of this deep uncertainty should be stamped into M, in the blacksmith's shop, so to speak, of the proper and rigorous process of identification. Like the red flag signalling danger and insecurity, when called upon (if it is) in the subsequent problem solving of 'Given $[u, M]$ find y!', it should manifest itself in a ballooning of the uncertainty surrounding the future forecast response ($y(t^+)$) to any contemplated environmental policy.[105]

Whatever else might be claimed, our collective instinct is to build complex simulation models M *precisely* for this purpose of generating insights – in the here and now, to shape and implement policy *now* – into the nature of that radically different future behaviour.[106] The inherent inadequacy, if not impotence, of the past data ($[u, y; t^-]$) will be manifest in their being unable to 'speak for themselves' about any such radically different future.

This is problematic, because (first) evidence of the deep uncertainty may not be apparent in the statistics conventionally computed for model evaluation.[107] It may not be apparent, because it is not sought out, since it is not believed to exist: witness the prevalence of model calibration and the nigh-on historical neglect of the error-driven process of discovery (with M) embedded in model structure identification;[108] as confirmed by some of the remarks in Reichert and Mieleitner[109] and Tomassini et al.[110] Alternatively, this deep uncertainty may not be apparent, because it *is* sought out, but proves so very hard to disentangle from the effects of everything else in the welter of resulting evidence. Technically, evidence of its presence may be revealed in the structure of the errors attaching to the posterior estimates of α, (after identification) and the more detailed properties of the mismatches of $[M \ vs \ `T`]$, typically properties below those captured in the customary headline statistical tests. But more broadly, rigorously reconciling large models with big data, with the express purpose of learning and discovery, is far distant from being a straightforward algorithmic and procedural problem to solve. Experience shows, too, that the availability of big data can render untenable a previous position: of the mismatch between the large M and (previously) sparse data being attributed to that very sparseness.

This state of affairs – the inability to expose and address the issues of deep uncertainty in the model – prevails because of the ease with which (technically) any number of variations on the theme of structure M (and the values of its parameters α) may fit the past data equally well, and quite possibly misleadingly well. This will be all the more so, the larger the model M, in particular, when not all of its parameters can be considered 'universal constants', hence the principle in favour of shunning calibration on the part of air quality modellers (as above; NRC; and also Box 1). The superficiality about this capacity to fit the model to the data at will has probably contributed to some of the distrust of models and the disrespect of, or distaste for, modellers. In detached, professional terms, the problem is referred to as a lack of model identifiability, something of which we have been more than fully aware since the 1970s (and well before, in fact).[111] In more colourful language, with doubtless greater popular immediacy (albeit rather threatening), it has been called 'equi-finality'.[112] This is why some would like to have, first, models based solely on the laws of physics bearing parameters – indeed *constants*, as though fixed points in an otherwise ever fluid knowledge base – and, second, models that do not require calibration against the past field data (as discussed further in Box 1). The grandness of this 'grand culmination' in modelling (as noted on pp. 108–17), should now be more apparent.

Deep uncertainty in the model M, given the desire to explore futures that may be radically different from the past is problematic, because even if the evidence of the deep uncertainty (as defined on p. 102 as MSEU) has been accounted for in respect of system identification, some of it – the very deepest of MSEU – is technically very, very difficult to account for in its propagation into the forecasts of the future, i.e. in generating high-quality, albeit deeply uncertain, future data ($[u, y; t^+]$). In general, this deepest kind of uncertainty has not therefore been accounted for.[113]

What is the material consequence of all this for handling uncertainty in models at the science–policy-society interfaces?

Problem Solving Across the Life Cycle of a Policy

It is a poor author, perhaps, who can write only about problems and challenges and not respond to such a question. Moreover, all the apparatus of our technical notation and algebra has now been set up. It would be a shame not to use it for outlining solutions to some of the problems, or reframing them in novel ways (and then solving these reframed problems). To enter into any detail is, of course, well beyond the scope of this chapter; nonetheless, wherever possible the reader is given pointers to other publications containing such detail. Yet we still need our algebra for precision and specificity, without (we hope) clogging the flow of ideas with notation and quasi-mathematical expressions.

A Different Kind of Foresight

Suppose someone, somewhere, perceives an oncoming threat to the environment. Do we need a policy to meet it, in anticipation?

What most interests us, as we have now seen, is the deep uncertainty surrounding whether future behaviour of the environment will be essentially similar to that of the past or radically different from it. Expressed more tightly, under more of the same, $[u, y; t^+]$ approximately equals $(\approx)$ $[u, y; t^-]$. Under profound change, $[u, y; t^+]$ decidedly does not equal $(\neq)$ $[u, y; t^-]$. In the past (the 1960s), merely the scientist would have offered his/her opinion (O1) on what might happen in the future, just as Morgan[114] describes such reasoning, with what she understands as the simple models of economics (not the complex computational models M at the focus of this chapter; see also Box 1 and Mari and Giordani (this volume)).

For the purpose of generating insights in the here and now, to shape and implement policy in the *present* (t_p), neither past (t^-) nor future (t^+), we have been experimenting with procedures for solving a rephrasing of an 'inverse' problem, as follows.[115]

Given some conjectures about future data, or 'the future', ideally *not* generated by any M (as we shall see), identify those parts of the model (technically, values of its parameters, α) that are key to discriminating between whether any one of those conjectured futures is 'reachable' (attainable), or not. Expressed in our algebraic notation, we wish to cleave those parameters so identified as key (α_{key}) from those that are not, i.e. parameters that are redundant to fulfilling the specified task, $(\alpha_{redundant})$, where such separation is driven by the nature of this *predictive* task. Given that these 'parts' in the model (all the α) can be composed as representing elements of policy, technology and the science base, we have a means of uncovering key elements of this 'policy, technology and the science base' relative to the reachability or otherwise of some desired (or feared) future. In other words, among the multitude of possible knowns and unknowns in the science base – the plausibles of the {the presumed known} and the implausibles of the {acknowledged unknown} assembled within M – we wish to identify just the handful of those most critical things, relative to the conjectured future, that we need to do (i.e. policy), innovate (i.e. technology) or understand better (about the science), with less uncertainty, and as soon as possible, as a priority in the present.

Put formally and cryptically, this re-phrased inverse problem is:

'Given $[u, y; t^+]$ and M, find α_{key} and its complement $\alpha_{redundant}$!'.

This would be tantamount to calibration of the model against some expression of the future, *but* a future that is anything but generated by that self-same model. In the best traditions of the scientific method, the contents of $[u, y; t^+]$ and those of *M* should be maximally distinct experiences of the way the world behaves.[116]

Our contemporary context for problem solving is that of extended peer review and participatory policymaking, i.e. what we have called the 'complex' at the science–policy–society interfaces. Furthermore, as we have also said, any singular consensus about the future is anathema to foresight generation. Of particular interest to us, therefore, is the way in which this alternative means of generating foresight would celebrate the richness of all the multiple conjectures, or opinions O_i, of each of the stakeholder groups $i = 1, 2 \dots m$, especially those at the periphery of our field of vision. That is to say:

> Given all manner of orthodox, rebellious, cranky and other (even bizarre) conjectures about future behaviour $[u, y; t^+]i$ derived from all the O_i; given possibly deeply uncertain, yet mathematically specific and explicit, conjectures about the content of *M*; find those α_{key} (and $\alpha_{redundant}$) upon which the reachability of each and every one of the multiple $[u, y; t^+]_i$ turns!

Of possibly great appeal, any policy elements of the α_{key} found to be key across the board, as it were, might be termed (desirably) non-foreclosing actions. They would be actions to be taken now that would not foreclose – for the time being – on the reachability of any one (i) of the stakeholders' passionately held views on what constitutes a desired, distant future $[u, y; t^+]_i$, across the board of $i = 1, 2 \dots m$. Forms of these problem-solving procedures are already being used to explore issues of technological innovations and governance for enabling cities to become forces for good in their environments.[117]

It is time to recall the essence of why this chapter has been written. The deep uncertainty in the science base (see pp. 101–7) should be reflected here in the content of *M*. The deep uncertainty of the contradictory certainties of contending convictions of the way the world is (see pp. 108–17), should be manifest now in the plurality of aspirations (y) entailed in all the $[u, y; t^+]_i$. To best serve scientific enquiry, anyone furnishing content for *M* should *not* be an author (i) of any future pattern of the system's behaviour. For the pragmatic purposes of computation, the aspirations (y) might need to be even grossly uncertain in numerical terms, yet nevertheless preserve faithfully the qualitative, contradictory certainties of the oral-verbal expressions of each set of such convictions, which each decision-stakeholder resolutely maintains. The approach and algorithms of Osidele and Beck[118] are indicative of the means of doing so.

Structural Change in the Behaviour of the System

Nothing is constant, as we have already observed (on pp. 101–7). In the very long run, evolution works to generate new, freestanding species. In the shorter term, non-native species can be introduced into, or invade, an existing ecosystem – much as the advent of models and modellers has disrupted the established order and structure of the interactions within the prior scientific community (see pp. 117–26). The structure of the system afterwards is not as it was before; the way its parts are pinned together, hence interact, cannot be as before, because there are entirely *novel* parts within it. This is a metaphor for establishing a principle: that the structure of the model M for time past (t^-) may be different from that for time future (t^+), i.e. in general terms, we may view the nature of the model as a function of time t, i.e. $M(t)$, as we know, but in a manner now amplifying the cursory, introductory discussion of the second section of this chapter.

If, then, there is a suspicion of behaviour coming to be profoundly different in the future, i.e. $[u, y; t^+] \neq [u, y; t^-]$, it is possible this could originate in the deepest uncertainty of the past–present science base, i.e. that $M(t^-) \neq M(t^+)$, where past–present is signalled simply by t^-. In words, there can be structural change in the behaviour of the system and the content of the model. If this is so, two related questions arise, to which provisional responses have been begun, as follows:

> Given past $[u, y; t^-]$, and possibly some 'priming' conjectures on the (radically different) future $[u, y; t^+]$ towards which behaviour is destined or inclined (we believe), how might we detect – within what we have observed empirically, i.e. $[u, y; t^-]$ – the seeds of such structural change: the very beginnings of some imminent change, of $M(t^-)$ embarking (in the here and now) on a path towards $M(t^+)$? In today's vernacular, can we detect the subtle signs of our approaching a 'tipping point'?[119]

> Is there any way we might (likewise) identify from $[u, y; t^-]$ the implied deep uncertainty of $M(t^-) \neq M(t^+)$, characterize it in the simplest of mathematical forms – part *non*random bias and part random – and extrapolate the evolution of the *bias* (under uncertainty) into the future, hence to make more robust the assessment of contemplated policy (u^*) intended to steer our affairs (for the moment) away from the tipping point?[120] Technically, extrapolation of the bias is known as a 'generalized random walk' model; figuratively, it is the generally forward motion of the drunk.

In reflecting on this latter, one might ask: why not resolve the bias explicitly? This, however, would be akin to solving the problem of model structure identification, i.e. transforming these particular aspects of deep uncertainty into more shallow forms. This can take much time. Implementation of policy might not have such a luxury. The history of pondering the error-driven process of model structure identification (as knowledge discovery) recorded by Beck, Lin and Stigter[121] covers four decades. What is more, its end-point is a better $M(t^-)$, as a representation of past behaviour, not necessarily any good account of $M(t^+)$.

But let us proceed now to a more mature stage in the life cycle of a generic policy, beyond threat identification, or assessment of the need and scope of possible counter-measures.

Evaluation: To Trust or Not the Validity of the Model for the Given Policy Task

Consider an agency charged with protecting the environment and society from the possible harms of toxic substances propagating around that environment. Consider further man's creativity in synthesizing a myriad novel chemicals and genetically modified organisms – all ostensibly to serve beneficial purposes (for mankind) – but *never* encountered in the world before. A new component part is about to be introduced into the system. Under the already existing law (on environmental protection), should manufacturer C be given a permit to proceed to full-scale production, hence commercial release of its novel substance Z?

Future behaviour of the substance in the environment cannot have been observed in the past. Just as for the foregoing problem of generating foresight under structural change, technically, $[u, y; t^+] \neq [u, y; t^-]$ and $M(t^-) \neq M(t^+)$, in principle. In the 2010s, models (M) are being used to predict the fate of novel substance Z and they are the subjects of legal disputes, as already noted on pp. 108–17.[122] Should we place our trust in M, therefore, as the basis for a ruling, an actionable interpretation of a regulatory policy, on whether Z should be produced and released?

Before answering any of these questions, let us conceive of models as tools, designed to fulfil specified tasks:[123] models, that is, as succinct *archives* of knowledge; as *vehicles* for the discovery of our ignorance, and at the earliest possible juncture; as *devices* for communicating science to policymakers and members of society; as *instruments* of prediction; even as truth-generating *machines*. Consider, in particular, the task of attaining (or not) a future pattern of behaviour, $[u, y; t^+]$.

According to the conventions consolidated in the NRC book on *Models in Environmental Regulatory Decision Making*, assessing the trustworthiness of a model (model evaluation) in the above generic decision context – regarding the release of novel substance Z – must necessarily be confined to the body of evidence (E2) on the extent to which the model (M) mimics the past behaviour ($[u, y; t^-]$) of the real thing, i.e., evidence $[M \, vs \, 'T']$. Yet if, by definition, the behaviour of greatest interest is so manifestly *not* $[u, y; t^-]$, but $[u, y; t^+]$, can something of relevance be found in solving the reframed inverse problem of pp. 127–8, above: 'Given $[u, y; t^+]$ and M, find α_{key} and its complement $\alpha_{redundant}$!'? Could there be evidence to be obtained from evaluating the model against some expression of the future – some expression of its predictive task? Could we gauge whether a given model (M) is well- or ill-suited to predicting a 'high end' exposure of the population to substance Z? Could we gauge whether candidate model M_A or candidate model M_B is the better suited to this task? Is model M_A better or worse suited to the task of predicting extreme exposures or that of predicting average exposure levels?

We have experimented[124] with using some measure of the statistical properties associated the numbers of α_{key} and $\alpha_{redundant}$, such as some form of ratio, $\alpha_{key}/\alpha_{redundant}$, for example. Technically, we would probably be inclined to place more trust in a model whose α_{key} are known with less uncertainty than one with highly uncertain critical parameters. Consider, by way of analogy, placing one's trust in an engineering structure whose key structural members are relatively secure/insecure. Less clear cut, perhaps, is whether we would prefer to trust a model in which 3 of its total of 10 parameters are key to the task at hand, as opposed to just 1 of the 10? Would we trust equally well a model with 7 of its 10 parameters found to be key, as opposed to another model with 70 of its 100 parameters being key? Such thinking about this challenge draws us toward highly complex (and convoluted) notions of *quality* in the design of a tool to serve a specified purpose.[125] For the quality of tool-instrument design is in the eye of the beholder. Think of the hammer adorned with baroque ornament (plenty of $\alpha_{redundant}$) and one unadorned (few $\alpha_{redundant}$). Which would user (i) prefer to wield over the task of driving home the nail?

With every step we take in such a direction, the social and the subjective elements of handling deep uncertainty in models at the science–policy–society interfaces become ever more labyrinthine. And there is yet more to this line of thinking, as we progress to yet later stages in the life cycle of an environmental policy: the negotiations involved in a provisional policy/ruling transmogrifying into a final policy/ruling.

In a Court of Law: Structuring the Games People Play

We are accustomed to evidence on [M *vs* 'T'] categorized by what we shall call the *fidelity* of the model, for want of a better word that has not yet been appropriated for serving some similar technical purpose. The foregoing suggests now that some of the numerical evidence to be taken into account, whether technically a part of [M *vs* 'T'] or not, can have to do with a utilitarian property of *relevance* – to the task at hand. Evidence [M *vs* 'T'], however, will also have dimensions attaching to it with a bearing more on the *transparency* of the model. In giving evidence to the NRC Committee, as it prepared its book *Models in Environmental Regulatory Decision Making*, a member of a non-governmental, community activist organization (N, say) observed that her cash-strapped employer did not have the luxury of access to the technical expertise with which to come to its *own* view on the trustworthiness of the model (based on the conventional evidence of evaluation and scientific peer review). Her organization would have no capacity to 'speak for itself'. She worried, therefore, about the transparency of the model: what mischief might be made by others – the government (G) or the regulated (R), for instance – through manipulation of the opaque inner workings of the model?

What stance, in fact, might each decision-stakeholder come to on the categories of evidence, *fidelity*, *relevance*, and *transparency*? Based on the legal studies of Wagner et al. and Fisher et al,[126] with some judicious exaggeration and caricature, one can imagine this. G might be for high *fidelity*, very high *relevance*, but low *transparency*, in the interest of stifling legal challenges to its model (M_G). R would also be for high *fidelity*, very high *relevance*, but low *transparency*, yet with reference exclusively to *its* model M_R of the way the world works, where M_R is self-evidently and indisputably far superior in all these respects to the thoroughly uncertain model (M_G) of the government. Poor N would be for high *fidelity*, high *relevance*, but (above all) very high *transparency*. Indeed, N might well wish a plague on both the houses of the government (G) and the regulated (R). For the present legal scene is not encouraging from N's perspective, as described by Wagner et al.:[127]

> In sum, because they are contingent and technically complex, and yet at the same time enter a policymaking world that is not well prepared to use them wisely, models are fodder for abuse and manipulation.
> [...]
> Many rulemakings have significant economic implications for one or more affected industries [the R, as we have called them], and strategic actors face incentives to exploit the misunderstanding of models as 'answer machines' to advance their own narrow ends.
> [...]
> In an ideal world, strategic efforts by stakeholders to hijack models to suit their interests would be fended off by stakeholders of the opposing stripe.

Community activist, non-governmental organization N should indeed see itself as 'striped' decidedly differently from the wealthy R, whom they might well accuse of some hegemonic hijacking of the legal discourse. Access to the debate by these 'unrepresented sets of stakeholders' (the N) 'might be subsidized',[128] so that their voice may be heard – not drowned out by the shouting of the R – and merit even a fully reasoned response, from both these regulated parties (the R) and the government (G) alike.[129]

In short, it is presumably the purpose of a court of law to create and impose a strict structure for arriving at a singular outcome, i.e. policy action u^* (let us say), from the multitude of heterogenous players to which it grants standing; from all the evidence ((E1) through (E3)) it deems admissible, which evidence base may seem to be expanding with every sentence here being written; and from all the opinion ((O1) through (O3)), where now 'expert witness' is becoming more difficult to define, if such definition is important for the court in privileging some opinions over those of others.

How much more straightforward must life have been in the 1960s, when science alone spoke truth (O1) to power.

Adaptive Community Learning

The court scene we have just imagined is strongly redolent of decision making under contradictory certainties, the DMUCC of pp. 114–16. Within and beyond the court, towards thus maturity in the life cycle of an environmental policy (u^*), we have now two last questions to ask: how might the assembly and deployment of an M contribute in any way to easing the making of better decisions under such contradictory certainties? What, incidentally, might bestow the marque of distinction (*) on the u^*, i.e. how is 'better' to be defined?

In one of his seminal pieces of work, Holling[130] defines and illustrates a principle of adaptive management that has come to dominate contemporary discussions of stewardship of the man–environment relationship, albeit often not as Holling originally intended. Like the engineering of guidance systems for rockets and missiles, Holling's adaptive management was to fashion and use u^* as an action that would be an exquisite combination of both steering the man–environment relationship in some desired direction *and* probing the behaviour of the environment at one and the same time – to reduce the uncertainty in our understanding of it. The presumption was that the natural environment behaves as though an $M(t)$, a model with changing structure. Policy u^* would be adapted from time to time, based on the current snapshot ($t = t_p$) of what constitutes understanding of the behaviour of the system, again and again – a road without end, always learning, never getting it right.[131] Decisions, policies and rulemakings are *not* made 'once and for all'.[132]

If this then was policy u^* for reducing uncertainty in the natural science (of the forms discussed on pp. 101–7), can we conceive of shaping policies today with the express, complementary purpose (as we have now enquired) of probing the man–environment relationship for learning about the 'man' part of this relationship (as on pp. 108–17 and pp. 117–26)?

In nascent, rudimentary form the procedure sought has been called 'Adaptive Community Learning'[133] and in that form it has indeed made use of the kind of computational M at the core of this chapter.[134] As has since become apparent, there are worse and better ways of easing arrival at achieving the qualification (*) in the expression of policy u^* from the 'mess' and the deep uncertainty of conflict and disagreement by which DMUCC is defined. It is the refurbishment of Dahl's pluralist democracy, as noted on pp. 113–14.[135] How it might work in shaping policy and technological innovation for cities, such that they may become forces for good in their environment, is presented elsewhere.[136] No use of any M is necessarily entailed in this institutional process – called 'clumsiness' to jar thinking out of the elegant optimizing mind-set of mainstream DMUU (decision-making under uncertainty). How this might come about, however, has been aired in Beck elsewhere,[137] using the algebraic notational conventions of the present chapter.

Conclusions

Looking back half a century, to the beginning of the 1960s, there was none of this. Even the phrase 'speaking truth to power' had barely first been uttered. Man's landing on the moon was imminent, however. And that marked the point (1969) when the new, computer-based methods of 'systems analysis' began to flow with increasing volume, out from the aerospace industry and into just about every other aspect of our lives and decision-making therein. The guiding of environmental policy was no exception. We would not have recognized the ambition (at the time) of achieving a 'predictive science of the biosphere'. But it was somehow in the air and in the youthful exuberance of that environmental systems analysis of the early 1970s. It was the tacit expectation. A brief period of computational determinism preceded the subsequent concern for uncertainty about the resulting mathematical models. Perhaps it is thus not so curious that, in some perverse, reverse order, we are today much exercised by *deep* uncertainty – in spite of forty years of developing and applying ever more complex computational models (*M*) of environmental systems.

Contemplating now climate change, and what to do about it, is inconceivable in the absence of very high order models – big models. They, and their reliability and uncertainties, are the central concern in Smith and Petersen (this volume). We are surrounded by 'big data', both as observed of the real world and as generated from the computer world with these big models. And yet, there remains this deep uncertainty to be handled at the interfaces among science, policy and society.

The nature of this deep uncertainty, as we have described it in this chapter, is twofold. In part it resides in the science bases mobilized in the models. And in part it arises from debate and disagreement among those scientists, policymakers, lobbying groups and members of the public, who assert they hold a stake in the decisions to be made – and which decisions are to be founded upon the big models and their forecasts. They are presented as two (of the thirteen) grand challenges for environmental modelling in *Grand Challenges of the Future for Environmental Modeling*.

We have not argued in this chapter that ever larger computer models will permit scientific uncertainty to be ground into an ever more insignificant dust. Even if this were the prospect, there is little evidence that consensus would break out among all the various parties with a stake in the shaping and the outcomes of environmental policies.[138] We have no reason to believe that games would cease to be played with models at the science–policy–society interfaces; that we would cease to assert that an opponent's model is utterly shot through with uncertainty, while yet ours is a sure-fire certainty – as a representation of the truth of the matter. And 'we' do so, so obviously, because 'our' 'truth of the matter' entails what 'we' want for the future.

There are two sources of insights into the ways in which we might handle this kind of deep, argumentative uncertainty in the future:

(I1)   From legal studies,[139] the recommendation is that, first, 'courts will in general be resolving disputes about whether a model complied with guidelines (for ensuring transparency of the methods and assumptions), rather than deciding whether a model is "good science" or "arbitrary" compared to an alternate model propounded by a litigant' and, second, that 'balanced participation to provide a check on strategic game playing' should be ensured,[140] in particular, when the risk is of this game being played by just one (powerful) stakeholder.

(I2)   From anthropology,[141] there is the means to discern how to move up from the lowest forms of contestation – termed 'closed hegemony', i.e., the situation of one overwhelmingly dominant stakeholder – towards a structure for the highest, i.e., a refurbished form of Dahl's pluralist democracy, termed 'clumsiness' (following, we note, the original use of the phrase 'clumsy institutions' by legal scholar Shapiro).[142]

Both (I1) and (I2) take continuous learning and adaptation as self-evidently integral to the process, as does Adaptive Community Learning.[143]

We have been greatly concerned with the 'futures' orientation of the subject of this chapter, specifically how to deal with a future in which the behaviour of an environmental system may be *radically different* from what has been witnessed (and cherished) in the past. Putting science into practice for this purpose, even after all these decades of ever larger computational models, remains beset with uncertainty, indeed deep uncertainty (possibly counter-intuitively). Could this be a function (in part) of our increasing awareness of the ever-increasing volume of the glorious detail of the inner workings of spaceship Earth, so that there is just so much more to fear about the future? In retrospect, it has to be said, the book *Environmental Foresight and Models: A Manifesto* was driven by this sense of future 'threat'; a focus more on what we fear rather than what we desire. For the science we put into practice, therefore, we would now see this as a matter of designing and employing models (*M*) expressly as vehicles for the task of discovering our ignorance, if not error and uncertainty – and as soon as possible.

6 VARIATIONS ON RELIABILITY: CONNECTING CLIMATE PREDICTIONS TO CLIMATE POLICY

Leonard A. Smith and Arthur C. Petersen[1]

Introduction

This chapter deals with the implications of uncertainty in the practice of climate modelling for communicating model-based findings to decision-makers, particularly high-resolution predictions[2] intended to inform decision-making on adaptation to climate change. Our general claim is that methodological reflections on uncertainty in scientific practices should provide guidance on how their results can be used more responsibly in decision support. In the case of decisions that need to be made to adapt to climate change, societal actors, both public and private, are confronted with deep uncertainty. In fact, it has been argued that some of the questions these actors may ask 'cannot be answered by science'.[3] In this chapter, the notions of 'reliability' are examined critically; in particular the manner(s) in which the reliability of climate model findings pertaining to model-based high-resolution climate predictions is communicated. A broader discussion of these issues can be found in the chapter by Beck, in this volume.

Findings can be considered 'reliable' in many different ways. Often only a statistical notion of reliability is implied, but in this chapter we consider wider variations on the meaning of 'reliability', some more relevant to decision support than the mere uncertainty in a particular calculation. We distinguish between three dimensions of 'reliability' (see 'Three Notions of Reliability', pp. 140–5) – *statistical reliability*, *methodological reliability* and *public reliability* – and we furthermore understand reliability as reliability *for a given purpose*, which is why we refer to the reliability of particular findings and not to the reliability of a model, or set of models, per se.

At times, the statistical notion of reliability, or 'statistical uncertainty', dominates uncertainty communication. One must, however, seriously question whether the statistical uncertainty adequately captures the 'relevant dominant uncertainty' (RDU).[4] The RDU can be thought of as the most likely known unknown limiting our ability to make a more informative (perhaps narrower, perhaps wider, perhaps displaced) scientific probability distribution on some outcome of interest; perhaps preventing even the provision of a robust statement of subjective probabilities altogether. Here we are particularly interested in the RDU in simulation studies, especially in cases where the phenomena contributing to that uncertainty are neither sampled explicitly nor reflected in the probability distributions provided to those who frame policy or those who make decisions. For the understanding, characterization and communication of uncertainty to be 'sufficient' in the context of decision-making we argue that the RDU should be clearly noted. Ideally the probability that a given characterization of uncertainty will prove misleading to decision-makers should be provided explicitly.[5]

Science tends to focus on uncertainties that can be quantified today, ideally reduced today. But a detailed probability density function (PDF) of the likely amount of fuel an aircraft would require to cross the Atlantic is of limited value to the pilot if in fact there is a good chance that metal fatigue will result in the wings separating from the fuselage. Indeed, focusing on the ability to carry enough fuel is a distraction when the integrity of the plane is thought to be at risk. The RDU[6] is the uncertainty most likely to alter the decision-maker's conclusions given the evidence, while the scientist's focus is understandably on some detailed component of the big picture. How can one motivate the scientist to communicate the extent to which her detailed contribution has both quantified the uncertainty under the assumption that the RDU is of no consequence, and also provided an idea of the timescales, impact and probability of the potential effects of the RDU? The discussion on the main case analysed in this chapter suggests that failure to communicate the relevant 'weakest link' is sometimes under-appreciated as a critical element of science-based policymaking.

Arguably the Intergovernmental Panel on Climate Change (IPCC) has paid too little attention to communicating the RDU in simulation studies, even though within climate science, and particularly within the IPCC, there has been increased attention on dealing with uncertainty in climate models over the last fifteen years or so.[7] It remains questionable whether this increased attention has led to a sufficient understanding, characterization and communication of uncertainty in model-based findings shared with decision-makers. Early model studies were often explicit that the quantitative results were not to be taken too seriously.

While the IPCC has led the climate science community in codifying uncertainty characterization, it has paid much less attention to specifying the RDU. The focus is at times more on ensuring reproducibility of computation than on relevance (fidelity) to the earth's climate system, in fact it is not always easy to distinguish which of these two are being discussed. Instead, the attention has mainly been on increasing the transparency of the IPCC's characterization of uncertainty. For instance, the latest IPCC guidance note for lead authors who are involved in the writing of the Fifth Assessment Report,[8] which was endorsed by the IPCC governments at its 33rd session (10–13 May 2011, Abu Dhabi), like its two precursors emphasizes that uncertainties need to be communicated carefully, using calibrated language for key findings, and that lead authors should provide traceable accounts describing their evaluations of evidence and agreement. While the basic message of the latest IPCC guidance thus does not differ significantly from the first,[9] the IPCC has learned from past experience and from the recent evaluation by the InterAcademy Council,[10] and has made more clear when to use which calibrated uncertainty terminology. Also, given the turmoil surrounding the IPCC in 2010, it is understandable that its lead authors have been asked to put more effort into providing traceable accounts of the main findings and their uncertainty qualifications in the Fifth Assessment Report (AR5), which was due for publication in 2013 and 2014. Such accounts were largely lacking in the past assessment rounds.[11] Still, being transparent, while certainly a good thing in itself, is not the same as communicating the RDU for the main findings.

The IPCC is not the only effort climate scientists are engaged in to assess and communicate findings on future climate change to decision-makers. In countries all over the world, ongoing projects aim to tailor information from climate models for use by decision-makers. These projects and their dissemination ultimately feed back into the IPCC assessment process, which periodically assesses the (peer-reviewed) literature in the context of climate science. In this chapter we critically reflect on one particular project which ran in the UK. This UK project produced the United Kingdom Climate Projections 2009 (UKCP09) and exemplifies perhaps the largest real-world case to date of climate decision support at very high resolution (post-code or zip-code resolution through the end of this century) based upon climate models, within a so-called 'Bayesian' framework. A great deal has been learned about error, insight and decision-making in the course of this groundbreaking project; one of the aims of this chapter is to explore how insights gained in this project can be used to minimize future confusion, misunderstanding and errors of misuse, avoiding spurious precision in the probabilistic products while maintaining engagement with user communities.

A central insight is to note that when the level of scientific understanding is low, ruling out aspects of uncertainty in a phenomenon without commenting on less well understood aspects of the same phenomenon can ultimately undermine the general trust decision-makers place in scientists (and thus lower the public reliability of their findings). Often epistemic uncertainty or mathematical intractability means that there is no strong evidence that a particular impact will occur; simultaneously there may be good scientific reason to believe the probability of some significant impact is nontrivial, say, greater than 1 in 200. How do we stimulate insightful discussions of things we can neither rule out nor rule in with precision, but which would have significant impact were they to occur? How do we avoid obscuring the importance of things we cannot rule out by placing (burying) the warning on an agreed high impact RDU in a long list of standard, inescapable caveats? How can those throughout the evidence chain from science through modelling to analysis, consulting and ultimately decision-making, be incentivized to stress the weakest links in their contributions and in the chain itself? Doing so would allow improved decision-making[12] both in the present and in the future. This chapter addresses these questions and is structured as follows.

On pp. 140–5 we introduce the three variations on the meaning of reliability. In the next section entitled 'Reliability$_2$ and Climate-Model Based Probability Distributions' on pp. 145–8, model-based probabilities are cast within a Bayesian framework,[13] stressing the importance of the 'information' *I*, the evidence assumed to be effectively true in order to convey any decision-relevance to the probability distributions generated; the extent to which climate-model derived probability distributions can be considered methodologically reliable is put into question. 'A Close Look at UKCP09: Definitions and Scope' on pp. 148–52 notes the need to resist shifting definitions from what counts to what can be counted today. This section focuses on the UKCP09 report, noting several passages and presentations which point to fundamental limitations in the decision-relevance of its outputs, but not in a manner likely to be accessed by users of its outputs. Examples are given where distinguishing obfuscations from clarifications of fact is nontrivial. The concluding section on pp. 153–6 returns to the question of error in science, of how uncertainty and progress in climate science differ from other decision-relevant sciences, including the challenges posed by an active anti-science lobby. Climate science can likely be most quickly advanced and most usefully employed when its errors, shortcomings and likely failures are laid as bare as can be. How are all actors in the chain – from science, through modelling, criticism, decision support to policymaking – best incentivized to use the information available today well and maximize the likely information available tomorrow? Recognizing and clearly identifying the RDU in each domain is a useful start. The Dutch approach towards climate scenarios is compared with the UK approach in passing, and argued to be more appropriate.

In short, this chapter addresses the question of whether the practice of climate modelling and communicating results to decision-makers suffers from shortcomings that may lead to problems at the interfaces between science, policy and society (see Beck, this volume).

Three Notions of Reliability

Climate science is confronted with 'deep uncertainty'.[14] This is true both when one uses climate models to investigate the dynamics of the climate system in order to generate understanding and when one uses them to predict particular (aspects of) future states of the climate system. Deep uncertainty in findings based on climate models derives from two distinct sources:

A There are fundamental limitations in our own cognitive and scientific abilities to understand and predict the climate system probabilistically, even when given the future pattern of emissions (which is a form of 'epistemic uncertainty').

B There are fundamental limitations to predictability inherent in the nature of the climate system itself and its surroundings (which refers to 'ontic uncertainty').[15]

A implies that:

1. The difference in the range of model outcomes under different models may be large.
2. One cannot justify the assignment of statistical meaning to a distribution of model outcomes in reference to the real world; this distribution can at best be regarded as a reflection of 'model scenario uncertainty' or a 'non-discountable envelope'.[16]
3. It is crucial to recognize and acknowledge ignorance and its implications: one knows that the models do not capture some dynamical features which are expected to become important in the context of climate-change studies. Additional features will play this role as a function of lead time. It is not clear what is the most reliable methodology for climate modelling (it is clear, though, that all climate models suffer from nontrivial methodological flaws, in terms of the limits of the technology of the day, including the favoured theoretical and empirical underpinning of assumptions, and similar hardware constraints).
4. Some model assumptions are strongly influenced by the particular epistemic values (general and discipline-bound) and non-epistemic values (socio-cultural and practical) which come into play in the design of both model and experiment.

And *B*, in combination with *A*, implies that:

5. Surprises are to be expected; unanticipated, perhaps inconceivable, transitions and events cannot be excluded.

The reader is referred to a book-length treatment of uncertainties in climate simulation for a more detailed explanation and underpinning of these claims and how they have played out in climate-policy debate to date.[17]

Additional reflection is due both in the practice of climate science and in the practice of assessing and communicating that science to decision-makers. Ultimately, uncertainty assessment and risk management rely upon human expert judgement. This makes it even more important in assessments to provide a traceable account of how expert judgements on uncertainty have been reached. And since the reliability of findings are contingent on the reliability of the selected experts themselves, establishing the overall reliability of these findings for decision-makers becomes an even more daunting task.

We here propose to distinguish between three dimensions of reliability:
1. statistical reliability (reliability$_1$),
2. methodological reliability (reliability$_2$), and
3. public reliability (reliability$_3$).

Each of these dimensions plays a role in the public debate about 'the' reliability of our quantitative vision of the climate in the future, including model-based high-resolution climate predictions. They deserve to be consciously dealt with by the community of climate scientists.

Statistical Reliability (Reliability$_1$)

A statistical uncertainty distribution, or statistical reliability (denoted by reliability$_1$), can be given for findings when uncertainty can be adequately expressed in statistical terms, e.g. as a range with associated probability (for example, uncertainty associated with modelled internal climate variability). Statistical uncertainty ranges based on varying real numbers associated with models constitute a dominant mode of describing uncertainty in science. One cannot immediately assume that the model relations involved offer adequate descriptions of the real system under study (or even that one has the correct model class), or that the observational data employed are representative of the target situation. Statistical uncertainty ranges based on parameter variation are problematic in climate science, however, since the questions regarding the best model structure for climate models and the extent to which we can access it today remain open, as is argued on pp. 145–8.

Both objective and subjective probabilities have been used in expressing reliability$_1$. Confusingly, sometimes combinations also occur. For instance, the IPCC has used a combination of objective and subjective probability in arriving at its main findings on the attribution of climate change to human influences.[18] Even more confusingly, different meanings of 'subjective probability' are often confounded, especially within competing Bayesian analyses. We point to the clear distinctions made by I. J. Good. Of particular importance is the distinction between the Bayesian's target (which is the subjective probability of an (infinite) rational org) from the assortment of other types of probability in hand.[19]

Methodological Reliability (Reliability₂)

There are both epistemological and practical problems with maintaining a strong focus on the statistical reliability (reliability$_1$) of model findings. We know that models are not perfect and never will be perfect. Especially when extrapolating models into the unknown, we wish 'both to use the most reliable model available and to have an idea of how reliable that model is,'[20] but the reliability of a model as a forecast of the real world in extrapolation cannot be established. There is no statistical fix here; one should not confuse the range of diverse outcomes across an ensemble of model simulations (projections), such as used by the IPCC, with a statistical measure of uncertainty in the behaviour of the earth. This does not remotely suggest that there is no information in the ensemble or that the model(s) is worthless, but it does imply that each dimension of 'reliability' needs to be assessed.[21]

And so another precise definition of 'reliability' results from consideration of the question: what is the referent and what is the purpose of a particular modelling exercise? If the referent is the real world (and not some universe of mathematical models) and the purpose is to generate findings about properties of the climate system or prediction of particular quantities, then 'reliability$_1$' is uninformative: one can have a reproducible, well-conditioned model distribution which is reliable$_1$ without reliable being read as informative regarding the real world.

Two distinct varieties of statistical uncertainty ranges are often estimated, one from comparing the simulation results with measurements – provided that accurate and sufficient measurements are available – another from sensitivity analysis within the model – provided that the accuracy of the different components in simulation practice (e.g. model parameters) are well-defined and known. Of course in climate-change modelling, one does not have verification of outcomes as one has in weather forecasting. Realizations of weather-like simulations can continuously be compared with measurements and predictive skill can be determined, which is not possible in climate simulation.

Models, when they do apply, will hold only in certain circumstances. We may, however, be able to identify shortcomings of our model even within the known circumstances and thereby increase our understanding.[22] As was observed in the previous subsection, a major limitation of the statistical definition of reliability is that it is often not possible to establish the accuracy of the results of a simulation or to quantitatively assess the impacts of different sources of uncertainty. Furthermore, disagreement (in distribution) between different modelling strategies would argue against the reliability$_2$ of some, if not all, of them. An alternative is therefore to define the reliability of findings based on climate models in more pragmatic terms. As Parker has shown,[23] in order to establish the 'adequacy-for-purpose' of a model in a particular context, scientists rely not simply on the statistical proximity of model results to, for instance, a historical dataset of the quantity of interest, but use a much more elaborate argumentation. This argumentation includes an assessment of the 'methodological reliability' of the model, for instance of the quality (fidelity) of the representation of a particular dynamic process that is thought to be of importance, for example, in modelling particular future changes.

So, one also holds *qualitative* judgements of the relevant procedures to derive findings on the basis of climate models, that is, of the methodological quality of the modelling exercise. A methodological definition of reliability, denoted by reliability$_2$, indicates the extent to which a given output of a simulation is expected to reflect its namesake (target) in reality. The nature of this reflection may be even further restricted in terms of the particular purpose for which those outputs are being employed, specifically its methodological quality. The methodological quality of a simulation exercise derives from the methodological quality of the different elements in simulation practice, *given the particular purpose to which the models under consideration are being put*. It depends, for example, not only on how adequately the theoretical understanding of the phenomena of interest is reflected in the model structure, but also, for instance, on the empirical basis of the model, the numerical algorithms, the procedures used for implementing the model in software, the statistical analysis of the output data, and so on.

It is rarely, if ever, a straightforward affair to determine the methodological quality of a finding, the extent to which the outcome of a scientific simulation will reflect future events in the real world. Ultimately, it may not be possible to do more than agree on the probability that the results of simulation will prove to be mis-informative.[24] Reliability$_2$ has a qualitative dimension, and the (variable) judgement and best practice of the scientific community provides only a reference for its extraction. It depends, for instance, on how broadly one construes the relevant scientific community and what one perceives as the purpose of the model. The broader the community, the more likely it is that the different epistemic values held by different groups of experts could influence the assessment of methodological quality. Criteria such as (1) theoretical basis, (2) empirical basis, (3) comparison with other simulations, (4) adequacy of the computational technology and (5) acceptance/support within and outside the direct peer community can be used for assessing and expressing the level of reliability$_2$.[25]

Public Reliability (Reliability$_3$)

In addition to the qualitative evaluation of the reliability of a model, increasingly the reliability of the modellers[26] is also taken into account in the internal and external evaluation of model results in climate science. We therefore introduce the notion of reliability$_3$ of findings based on climate models, which reflects the extent to which scientists in general and the modellers in particular are trusted by others.

The situation for climate scientists and climate modellers[27] concerning the use of high-resolution model-based predictions is different from that of, for instance, doctors and the 'placebo effect': in the latter situation an over-confident doctor may increase the chances that a placebo works. In a similar situation, climate scientists might choose to be 'overly humble' just as other physicists still tend to take care when discussing 'the discovery' of a Higgs-like Boson. As we argue below, climate scientists can indicate the shortcomings in the details of their modelling results, while making clear that solid basic science implies that significant risks exist. If climate scientists are seen as 'hiding' uncertainties, however, the public reliability (reliability$_3$) of their findings may decrease, and with it the reliability$_3$ of solid basic physical insight.

Assessment of climate-model simulation for decision-making aims for consensus on the risks faced along with a rough quantitative estimate of the likelihood of their impacts. Providing such assessment is in practice the fundamental aim of the IPCC[28] and does not imply agreement on the action or level of investment to make.

Reliability$_2$ and Climate-Model Based Probability Distributions

A good Bayesian does better than a non-Bayesian, but a bad Bayesian gets clobbered.[29]

Each and every probability distribution is conditioned on something. This claim led Sivia[30] to argue that one should always write (or at least think) of every probability P as the probability of x conditioned on I, $P(x|I)$, where I is all of 'the relevant background information at hand'. In a forecast situation x is the value of interest, say the temperature of the hottest day in Oxford in the summer of 2080, for instance. Sivia, with many other Bayesians, argues that absolute probabilities do not exist. Following Good[31] we would argue that it is often useful to speak as if absolute (physical) probabilities did exist, regardless of whether or not they do exist. Without omniscience, such probabilities may well not exist; with omniscience they may or they may not; as physicists we would note, following Mach,[32] that we will never be able to tell.

The challenges that face scientists, climate modellers and policymakers are not nearly so deep. Questions surrounding what is intended by I are nontrivial. UKCP09, for instance, appears to reinvent the expression 'evidence considered'[33] leaving unanswered the question of whether there was enlightening 'evidence suppressed'. It also provides no clear answer to the critical judgement call as to whether or not the current state of the scientific modelling is adequate for the purpose of forecasting x in the far future, full stop. We return to what was intended by the writers of UKCP09 below, the general point here is that if we are aware from the science that an RDU is not included within the evidence considered when constructing probability forecasts, then decision-makers have no rational foundation for using those probabilities as such. How can we motivate contractors, in academia and in industry, to lay bare the limitations of their analysis without the fear that (by respecting the limitations of today's science and simulations) they will be accused of failing to deliver on the contract?

The UKCP presentation introducing UKCP09 includes the statements 'The problem is we do not know which model version to believe' and 'The only way we have of assessing the quality of a climate model is to see how well they simulate the observed climate'.[34] The first statement might be taken to indicate that there is one model version which we might believe; no measure of how well these models simulate the current climate is given, but it is clear that some do better than others.

The decision-relevant climate simulations of any epoch must run significantly faster than real-time on the computers of that epoch. It is widely suggested within the UKCP09 guidance materials that relevant climate simulations of 2009 are available at postcode-level resolution over the remainder of this century. Some climate modellers suggest that this is indeed the case although no clear statement supporting the claims of the UKCP09 methodology appeared in the peer reviewed literature before this year (2013).[35] The importance of stressing the limitations on our ability to quantify uncertainty is stressed by Murphy et al.[36] and Stainforth et al.[37]

Consider a parallel case. Many of us could code up Newton's Laws of gravity and simulate the motion of the planets of our solar system. Suppose each of us did so. The models would use different integration schemes, computer chips, integration time steps, perhaps some of us would include Pluto and others would not. Some of us would make coding errors; others might not have access to an accurate snapshot of the positions of each planet, some of us might have accidentally tweaked our models in-sample so that we account for noise in the observations as well as signal. Nevertheless, assessing the quality of each run based on the observed motions of the past would allow one rank-ordering of the relative quality of each run. And looking at the diversity of forecasts from the 'better' models would, we believe, provide a reasonable distribution for the likely location of the earth for quite a long time into the future; for the planet earth, but not for the planet Mercury.

Mercury does not 'obey' Newton's Laws, at least not without adding non-existent planet(s) between Mercury and the Sun.[38] General relativity is required to model Mercury's orbit accurately. The fact that Mercury's orbit differed from expectations was known long before general relativity was available to allow realistic simulation. There are parallels here with climate, which include not knowing which calculation to make, and not being capable of accounting for effects we know are of importance in the calculation.

We could be unaware of the theory of general relativity; alternatively one might find that significantly fewer of us could code up the relativistic simulation than could code the Newtonian simulation. If all we have access to is our ensemble of Newtonian simulations, then the diversity of our simulations do not reflect our uncertainty in the future location of Mercury.[39] We can know that this is the case without being able to 'fix' it: the RDU in this case was structural model error demonstrated by the lack of fidelity in Newton's Laws applied to Mercury's orbit (we might, of course, think it to be some other unknown). Our model output for Mercury is not decision relevant regardless of how principled the statistical post-processing methods applied to the model ensemble may be. When the model class in hand is inadequate, and believed to be inadequate, the meaning of $P(x \mid I)$ is cast into doubt, as I is known to be incomplete, perhaps known to be false.

Let x be the location of Vulcan, which does not exist; let I be Newton's Laws and let {data} be all observations of the planets up to the year 2000. Would the Bayesian want to say that the integral of $P(x \mid \{\text{data}\}, I)$ over all space is equal to one? To zero? This integral is the often suppressed 'normalization constant' in the denominator of Bayes's Theorem. Its importance is noted by Sivia[40] in his Equations 1.10 and 2.2. If I is effectively inconsistent with the data, as observed when our model class is inadequate due to structural model error, and we either do not know how or cannot afford a better model class, the decision relevance of any model based probability statements is thrown into doubt, and with them the machinery of the Bayesians. It is not that Bayes's Theorem fails in any sense, it is that arguably the probability calculus (of which it is a fundamental part) does not apply.

And so it is with climate. We know of phenomena that are critical to climate impact in the UK which are not simulated realistically in the models that went into the UKCP09 probability distributions. Not merely ephemeral phenomena like clouds, but rather more solid phenomena like mountain ranges: the Andes, for example, are not simulated realistically. Thus to the extent that these phenomena would affect the derived probability distributions, these distributions should be considered unreliable (*reliability*$_2$), which is a qualitative judgement (see pp. 143–4). For instance, failure to realistically simulate 'blocking' was noted explicitly by the official Hoskins Review sponsored by the funders of UKCP09.[41] Due to this acknowledged shortcoming, there are many circumstances in which interpreting UKCP09 probabilities as decision-relevant probabilities would be a mistake with potentially maladaptive consequences. But how is information on this propagated along the chain from an international review complete with external reviewers' commentary to the decision-maker interested in how many consecutive clear days, or rainy days, or days above/below a given temperature are likely to be encountered in a typical year of the 2080s? Not well. We return to this below after noting that definitions might better be formed based on the underlying science than on what our current models might simulate.

A Close Look at UKCP09: Definitions and Scope

Climate was sometimes wrongly defined in the past as just 'average weather'.[42]

The definition of climate matters. Smith and Stainforth[43] argue that the temptation to take the limitations of today's models (or today's RDU) into account when defining terms should be resisted. Current definitions provided by the IPCC Glossary of 2007[44] and the American Meteorological Society Glossary of 2002 focus on averages, noting both variability and changes *in these averages*. Many older definitions, including the American Meteorological Society Glossary of 1959, discuss the 'synthesis of weather', however observed, embracing the entire distribution of higher dimensional weather states, and their time evolution, however expressed.

Defining climate as the distribution of such weather states allows a knowledge of climate to imply whatever averages may prove of interest to a scientist or a decision-maker. Defining climate in terms of average values and variances may make it more straightforward to evaluate climate models, while in reality removing the policy relevance of 'climate' by definition. Average values reflect neither the phenomena of interest to the physicist nor the impacts on individuals which determine policy.[45]

The earth's climate system appears sufficiently nonlinear that if state-of-the-art models are unable to realistically simulate the weather of a given epoch, then it is difficult to argue persuasively that they will get even the climatological averages of the following epoch correct. From either a scientific standpoint or a decision-making standpoint, the traditional definition of climate as a distribution is called for. Such a definition may imply that we cannot simulate even today's climate very well. In that case determining (1) why this is the case, identifying whether RDU is due to computational constraints, to incomplete theory, or to a lack of observations; (2) estimating the magnitude of that uncertainty as a function of spatial scales and time; and (3) providing an estimate of when we can expect to address that dominant uncertainty, are each of immediate value to the policy process. Knowing the time required before the current RDU is likely to be reduced to the extent that another head of RDU will take its place is of great value in weighing up the advantages of delaying action against its costs.

In discussion of simulations in the context of reality, it is of value to clearly separate the properties of 'model quantities' from the 'quantities' themselves. Manabe and Wetherald,[46] along with many other climate modelling papers of that time, repeatedly distinguish 'the state of the model atmosphere' from that of the atmosphere. Similarly, they often say the 'model troposphere', the 'model stratosphere', and so on. They note 'a mistake in the programming' and quantify it, suggesting its impact. And they conclude by noting boldly that 'because of the various simplifications of the model described above, it is not advisable to take too seriously the quantitative aspect of the results obtained in this study'. Manabe, Weatherald and other climate modellers took pains to clarify aspects of reliability$_2$ in their results, the focus being on understanding phenomena rather than quantifying events in the world.

Modern studies are rarely so blunt, nor so clear in terms of distinguishing model-variables from their physical counterparts. They often also lack clarity in terms of stating spatial and temporal scales (if any) on which the quantitative aspects of model simulations are not to be taken seriously, or on how those change as the simulations are run farther into the future. The disclaimers tend to state the obvious: that today's climate science cannot provide a complete picture of the future. It is, of course, doubtful that climate science ever will provide a complete picture inasmuch as climate prediction always remains extrapolation. Catch-all disclosures would be more valuable if accompanied by clear statements of specific known limitations of current insights regarding particular applications.

The UKCP09 Briefing Report[47] does contain some clear statements. For example, in a bullet point list on page 6 on the power of Bayesian statistical procedures: 'Errors in global climate model projections cannot be compensated by statistical procedures no matter how complex, and will be reflected in uncertainties at all scales.'[48]

This quote is taken almost verbatim from the Hoskins Review, although it might be misread to suggest that model errors will be reflected in the UKCP09 probability distributions ('uncertainties at all scales') which the Hoskins Review held that they will not. An online version of a similar bullet point list contains the statement:

> Models will never be able to exactly reproduce the real climate system; nevertheless there is enough similarity between current climate models and the real world to give us confidence that they provide plausible projections of future changes in climate (Annex 3).[49]

It is as true as it is uninformative to claim that models will never be able to 'exactly reproduce' their target systems. Do the authors of the report believe that the probabilities provided can be interpreted as the probabilities of events in the real world? The hyperlink provided in the report to 'confidence' above does not address this question, and in Annex 3 we find the following:

> We have no positive evidence that such factors would, if included, provide sources of uncertainty comparable with those included in UKCP09 (at least for projection time scales of a century or less), but this remains an issue for future research.[50]

This is followed by a discussion of assumptions 'imposed by limitations in computational resource' and that 'non-linear interactions between uncertainties in different components of the earth system are important at the global scale, but not at the regional scale, because our finite computing resources were not able' and then:

> we believe that the UKCP09 methodology represents the most systematic and comprehensive attempt yet to provide climate projections which combine the effects of key sources of uncertainty, are constrained by a set of observational metrics representative of widely-accepted tests of climate model performance, and provide a state-of-the-art basis for the assessment of risk, within limits of feasibility imposed by current modelling capability and computing facilities.[51]

We agree with this quote from Annex 3 in terms of what the UKCP09 probabilities are, but these clear caveats do not suggest to us that the UKCP09 probabilities can be taken to reflect what we expect in the 2080s, or as 'plausible projections of future changes in climate' at 25 km length scales across the UK.

In the original print version of the bullet points on page 6, the last point reads:

> The method developed by UKCP09 to convert climate model simulations into probabilistic estimate of future change necessitates a number of expert choices and assumptions, with the result that the probabilities we specify are themselves uncertain. We do know that our probabilistic estimates are robust to reasonable variations within these assumptions.[52]

This last sentence is perhaps key. While it is useful to know that the UKCP09 probabilistic estimates are robust to reasonable variations within these assumptions, it seems remiss not to acknowledge that those same estimates are known not to be robust to variations in (alternative) assumptions which are consistent with today's science, if that is indeed the case.

Annex 3 includes long discussion of small uncertainties and things that can be sampled and controlled for. It concludes:

> In summary, the UKCP09 projections should be seen as a comprehensive summary of possible climate futures consistent with understanding, models and resources available at present, but users should be aware that the projections could change in future, as the basis for climate prediction evolves over time.[53]

What appears to be missing in the entire (evolving) document is a statement as to whether or not *the projections are expected to change*. We would argue that they are not mature probabilities. We expect these probabilities to change without additional empirical evidence just as the Newtonian forecasts of Mercury were expected to change: not due to new science, but due to steps taken to address known shortcomings in the UKCP09 projections. Were that the case, the phrase 'evidence considered' takes on a Machiavellian hue.

To our knowledge, the UK Met Office Hadley Centre (MOHC) has made no claim that the UKCP09 probabilities are robust in terms of providing well-calibrated forecasts of future weather of the 2080s; UKMO support is notably lacking for the output of the weather generator which is central to many headline claims.[54] At the 2009 UKCP science meeting in the Royal Society, members of the UKMO repeatedly stated that they would not know how to improve them today, when asked directly if the probabilities were adequate for decision-making. Climate researchers are in a difficult position, particularly given the political climate into which they have been drawn. How might one incentivize scientists and consultants to open the can of worms regarding where a given climate report is known not to supply well-calibrated decision-relevant probabilities, rather than obscure the fact of the matter with discussions of 'best available' or our inability to do better in the near future? How do we communicate our uncertainty when probabilities conditioned on 'evidence considered' are expected to give a rosier picture of uncertainty than those conditioned on all the available evidence? Does the failure to do so not risk a backlash from numerate users when future reports make this obfuscation clear?

A scientist reading through the report will find what appears to be clear evidence that the UKCP09 probabilities should not be taken at face value as probability statements about the real world; red flags to a scientist might not be so obvious to non-scientists. A few are collected below; they suggest the user look at the detailed regional model results if 'variability is important to the individual user' (not that the output is adequate for purpose); they defend the Hadley Centre model as competitive with other models (not adequate for purpose); they claim to represent 'some aspects' of blocking 'with reasonable fidelity' (not claiming to be adequate either for the purpose of extracting weather impacts, nor for the purpose of guiding reasonable simulation of the evolution of the UK's geography as it changes over the six decades still to pass before 2080):

> Careful evaluation of such diagnostics from the RCM simulations and the weather generators is recommended in cases where such variability is important to the individual user.[55]
>
> It should be recalled from Annex 3, Figure A3.6, that current positions and strengths of the modelled storm track do not always agree well with observations, and this should be taken into account when assessing the credibility of their future projections. The HadCM3 ensemble shows a better agreement in present day location than most other climate models, and a reasonable agreement in strength.[56]
>
> The mechanisms for atmospheric blocking are only partially understood, but it is clear that there are complex motions, involving meso-scale atmospheric turbulence, and interactions that climate-resolution models may not be able to represent fully. The prediction of the intensity and duration of blocking events is one of the most difficult weather forecasting situations. The HadCM3 model does represent, with reasonable fidelity, some aspects of present-day atmospheric blocking in the N.

Atlantic region (see Figure A3.7) with the performance in summer better than that in winter. At other longitudes the model shows less fidelity, in particular in the Pacific sector. (An additional complication is that it is not clear that simply doubling the resolution of a climate model automatically produces a better simulation of blocking – in the case of one Met Office Hadley Centre model, this results in a degradation).[57]

The inconsistency of the three diagnostics makes it difficult to make a clear statement about the ability of the perturbed physics ensemble to simulate anticyclones, but in general the HadCM3 ensemble is competitive with other climate models.[58]

The insights gained in the construction of UKCP09 hold significant value for governments and decision-makers beyond its targeted areas in the United Kingdom. It is extremely valuable to learn that some things cannot be done now, nor in the near future. Knowing that some aspects of guidance will not be available in the near future is a great aid to decision-making. Making it clear that some of the known limits of today's models 'might' severely limit the relevance of the precise probabilities provided by UKCP09, would certainly be of value. But sometimes the report fails to give a clear 'Yes' even to such questions:

> The role of atmospheric blocking under climate change is currently a major topic of research. Might current model errors severely limit the reliability of climate change projections (e.g. Palmer et al. 2008; Scaife et al. 2008)? Might large changes in blocking, that current models cannot simulate, cause large changes in the frequency of occurrence of summer heat waves for example?[59]

Given this positive framing of the outputs of the project, how are other nations to learn of the severe limitations UKCP09 imposes, subtly, on the application of its outputs in practice? How would other nations debating similar studies evaluate the value of such a study, where it might have decision relevance useful within their borders, and where decisions must be made under deep uncertainty, that is, in ambiguity, without robust statements of probability? Failing to clarify the limitations of the exercise are likely to lead to misapplications within the UK, and potentially misguided attempts to obtain in other countries what the quotes above show has *not* been obtained for the UK.

How can those civil servants and politicians promoting UKCP09 from inside, as well as professional scientists, be incentivized to communicate their clear insights into its limitations? How can they resolve the conflict between attracting and engaging their target audience of users while laying bare the limited utility of the product currently available? The UK's goal to continue to make world-leading contributions towards the understanding of climate risks will be hampered if government projects like the Climate Change Risk Assessment adopt the naive interpretation of the UKCP09 outputs suggested by the report's headlines and fail to take on board the deep limitations and implication of the dominant uncertainties identified in the body of the report. How might future studies promote critical information on their own shortcomings in their headline results?

Conclusions: Insights and Improvement

You can't have a light without a dark to stick it in.
Arlo Guthrie

Like weather forecasting and medical diagnosis, climate modelling is riddled with error and inadequacy. That is not pessimistic, even if it is often suggested to be so; it is an opportunity for the sciences, not a reason to abandon science.[60] Medical diagnosis has the advantage that there are many, many people from which we can more quickly learn the behaviours of all but the most rare diseases. This aids our ability to advance the science and make better decisions regarding patient care. While long-standing philosophical arguments as to what (and how many) probabilities 'are' remain,[61] there is firm empirical evidence that they are useful; many variations[62] on the Bayesian paradigm have provided a valuable organizing framework for managing probabilities when we think we have them, and have improved the approach taken to a problem even when we do not. In the case of weather, we have only one planet to observe. Still we observe effectively independent weather events happening all over the world every day, and we have the opportunity to learn by observing systematic failures in our weather models day after day. Week after week, after week. Operational centres actively save observations corresponding to past forecast busts to re-examine with each future generation of weather models.

Climate modelling is different,[63] in that the lifetime of a model is less than the typical forecast time and that the forecast time is often longer than the span of a scientist's career. This makes it much more difficult to learn from our mistakes. Of course, the fact that climate science comes under attack from a politically motivated anti-science lobby complicates things, a point we will return to below. One scientific challenge to climate science is that it is not at all obvious we will ever be able to model climate at lead times of eighty years on length scales of neighbourhoods in any manner that is quantitatively informative to adaptation decisions. If true, merely knowing this is itself of great value to policymakers. Knowing the length-scales on which we can expect a 'Big Surprise' deviating from our model-based probabilities in eighty years, or forty years, or twenty years, requires a new approach to quantifying 'deep uncertainty'. Even only twenty years out, is the probability of a big surprise less than 1 in 200? What is this probability for 2050, given the known limits of our current models? How much more will we know in 2017? Such questions are of immediate value in terms of rational mitigation policy.

While the United Kingdom has pushed the envelope with the UKCP09 projections, the so-called 'Bayesian' framework adopted by UKCP09 was, even after its launch, unable to clarify the likelihood that its model-based probabilities will prove mal-adaptive as it looks farther and farther into the future. The probability distributions for the 2080s are presented as if they were as robust as those for the 2020s. Or our knowledge from observations of the 2000s which, of course, come with probability distributions due to observational uncertainties.

How do we move to a broader and deeper understanding of the implications of incomplete knowledge when we are denied the luxury of millions of experiments an hour? Stirling[64] stresses the importance of keeping it complex: of not simplifying advice under the pressure to make it useful or accessible. This advice may lead us beyond probabilities, and the ability to cope with instances when the denominator in Bayes's Theorem is zero (if, indeed, it is well-defined at all). But how do we do this?

One could never usefully refute the claims of long-range climate predictions empirically. We can, however, explore the strength of evidence (*all* the available evidence) for its likely relevance. The quotes given in the previous section show that the seeds of doubt are sown deep within the scientific text of the report, yet they are all but suppressed if not obfuscated in the headline sections. Every user[65] of the UKCP09 products can hardly be expected to know all the meteorological phenomena relevant to an application of interest, nor is it reasonable to expect individuals to examine 'diagnostics from the RCM simulations and the weather generators'. Known shortcomings of each generation of models should allow more informative information on the expected robustness and relevance of particular UKCP09 probability distributions as a function of lead time. Information on the relevant dominant uncertainty is more useful when it is identified clearly *as* the RDU; it is less useful when buried amongst information of other uncertainties that are well quantified, have small impacts, or are an inescapable fact of all scientific simulation. Given the understandable tendency of modellers to defend their models as the best of the current generation of models, and the difficulties of a team constructing a report cataloguing its shortcomings, perhaps an accompanying second scientific report, a 'minority opinion', is called for: an independent study clarifying the limits of robustness could be included in all such highly policy relevant scientific reports. Ideally scientific risk assessments will translate both the implications and the limitations of the science for decision-makers in each sector. Extrapolation problems like climate change can benefit from new insights into:how to better apply the current science, how to advertise its weaknesses and more clearly establish its limitations; all for the immediate improvement of decision support and the improvement of future studies. This might also aid the challenge of training not only the next generation of expert modellers, but also the next generation of scientists who can look at the physical system as a whole and successfully use the science to identify the likely candidates for future RDU. Finally, note that government sponsorship of these minority reports would not only aid immediate decision-making, but could also have the second order effect of tightening the claims and clarity of future primary reports.[66]

An alternative approach to using high-resolution climate predictions that has been developed in the Netherlands holds significant promise and seems likely to be more informative than the 2009 UK approach. We here refer to the approach taken by the Dutch government and its Royal Netherlands Meteorological Institute (KNMI) towards climate scenarios for the Netherlands.[67] The Dutch approach explicitly departs from the view that a full probabilistic approach to climate predictions at the regional scale is currently feasible. Recognizing the inherent predictability limits of climate and the different epistemic values held by different groups of experts, the Dutch climate experts argue for a different approach. For instance, with respect to changes in extreme weather events, narratives describing existing knowledge of the physics of extreme weather events accompanied by simulations of extreme weather events in well-calibrated numerical weather prediction models in present-day climate and a potential future climate setting are provided to give a realistic and physically consistent picture of both the types of events that need preparatory actions and the impacts of the adaptation decisions taken.[68]

Climate policy on mitigation and decision-making on adaptation provide a rich field of evidence on the use and abuse of science and scientific language. We have a deep ignorance of what detailed weather the future will hold, even as we have a strong scientific basis for the belief that anthropogenic gases will warm the surface of the planet significantly. It seems rational to hold the probability that this is the case far in excess of the '1 in 200' threshold which the financial sector is required to consider by law (regulation). Yet there is also an anti-science lobby which uses very scientific-sounding words and graphs to bash well-meaning science and state-of-the-art modelling. If the response to this onslaught is to 'circle the wagons' and lower the profile of discussion of scientific error in the current science, one places the very foundation of science-based policy at risk.

Failing to highlight the shortcomings of the current science will not only lead to poor decision-making, but is likely to generate a new generation of insightful academic sceptics, rightly sceptical of oversell, of any over-interpretation of statistical evidence, and of any unjustified faith in the relevance of model-based probabilities. Statisticians and physical scientists outside climate science (even those who specialize in processes central to weather and thus climate modelling) might become scientifically sceptical, sometimes wrongly sceptical, of the basic climate science in the face of unchecked oversell of model simulations. This mistrust will lead to a low assessment by these actors of the reliability$_3$ (public reliability) of findings from such simulations, even where the reliability$_1$ and reliability$_2$ are relatively high (e.g. with respect to the attribution of climate change to significant increases in atmospheric CO_2).

It is widely argued that we need decision-makers to think about climate more probabilistically, as we will never be able to tell them exactly what will happen in the far future. That argument does not justify the provision of probability distributions thought not to be robust. Indeed, numerate users already use probability distributions as such. As additional numerate users appear, what would be the impact of having to explain in 2017 that the probabilities of 2009 were still being presented as robust and mature in 2011 even when they were known not to be? How, in that case, would one encourage use of the 2017 'probabilities' (or those of 2027) to that same numerate audience? Learning to better deal with deep uncertainty (ambiguity) and known model inadequacy can significantly advance and foster the more effective use of model-based probabilities in the real world.

7 ORDER AND INDETERMINISM: AN INFO-GAP PERSPECTIVE

Yakov Ben-Haim

Introduction

Order and chaos have impressed themselves on consciousness throughout human history. Modern science attempts to uncover and understand orderliness in creation, leaving *tohu vavohu* for others to contemplate. In this chapter, we describe a conception of indeterminism – the unknown, the uncertain, the formless and void – that is relevant both to the scientific endeavour and to the practical attainment of reliable decisions in human affairs. *Tohu vavohu* can be characterized, even understood in some sense, without dispelling the mystery of the unknown.

Indeterminism – the lack of an orderly law-like progression of events – occurs both in the physical world and in human affairs. We will use info-gap theory to characterize this indeterminism and the responses to it.

The prototype of worldly indeterminism is quantum mechanics. Quantum mechanics has enjoyed more than a century of success, explaining black body radiation and the photo-electric effect in the early years, up to nuclear tunnelling, radioactive decay, anti-matter, diode and transistor physics, and more. And yet, quantum mechanics comes at a cost: constriction of the domain of scientific explanation, and weakening of the classical concept of causality. One grain of truth that quantum mechanics captures is that the physical world has an element of irreducible elusiveness, what one might call *natural* or *ontological indeterminism*.

Human affairs are full of surprises, and the most important are those that we bring upon ourselves. People make discoveries and inventions. People have discovered new continents, the structure of the atom, the size of the universe, the processes of biological evolution, and much more. People have invented the kindling of fire, the domestication of wheat, writing, the wheel, the mechanical clock, the printing press, electric motors, the internet, and so on. These inventions and discoveries alter the behaviour of individuals and societies in fundamental and far-reaching ways. However, tomorrow's discovery cannot be known today. We cannot know what is not yet discovered, so we cannot predict future behaviour in its entirety. The human world has an element of irreducible elusiveness, what one might call *human* or *epistemic indeterminism*.

Ontological and epistemic indeterminism share a basic attribute: they are unstructured and unbounded, they lack order and form and regularity. Rules or laws do not govern or generate patterns of indeterminism. The unknown is unconstrained even by what we know: knowledge, even if true of the past, need not hold in the future.

Info-gap theory provides a framework for understanding both ontological and epistemic indeterminism. We will discuss a conceptual framework for understanding the physical, ontological origin of quantum indeterminism. The info-gap concept of robust-satisficing underlies the probabilistic interpretation of the wave function. We will also explain how info-gap robust-satisficing describes epistemic indeterminism and supports responsible, reliable decisions in the face of epistemic indeterminism.

We begin with a generic conceptual discussion of info-gap decision theory in the next section. We describe ontological and epistemic indeterminism in the sections on pp. 164–7 and pp. 167–8. On pp. 168–70, we discuss the concepts of optimizing, satisficing and indeterminism, in preparation for employing the idea of info-gap robust-satisficing as a response to indeterminism on pp. 170–4.

Info-Gaps

In this section we describe the conceptual basis of info-gap decision theory.[1]

The Known and the Unknown

The known and the unknown form an exclusive and exhaustive dichotomy, but one that is highly asymmetrical. 'Knowledge' is to the 'unknown' as 'banana' is to 'non-banana' or 'linear system' is to 'non-linear system'. The asymmetry is in both size and quality. For instance, biologists estimate that the number of species that have not yet been identified substantially exceeds the number that have,[2] and the ratio of unknown to known is even greater if one includes extinct species. Our immediate physical world – the earth, the solar system – is known in far greater detail than the vast expanses of the universe with its countless stars and planets. The realm of the known is tiny compared to what is not known. The qualitative difference between the known and the unknown is also substantial, as we will see.

'Whereof one cannot speak, thereof one must be silent'.[3] However, from both practical and speculative points of view, the unknown is not the realm of silence. We can and must say very much about the unknown. I will be discussing the unknown as an epistemic entity, rather than an ontological one. In this subsection I will use anthropomorphisms and sometimes even metaphors, but in 'Info-Gap Robust-Satisficing and the Response to Indeterminism' (pp. 170–4) we will extend our discussion beyond exclusively human knowledge and ignorance and attempt to be more precise.

One of the first things to say about the unknown is that its boundaries, if they even exist, are unknown. For example, in the analysis of risk in many fields we don't know which bad event (if any) will occur, and we often cannot realistically specify a worse case. The inventiveness – pernicious or propitious – of the unknown future is boundless. Or, we don't know how many animal phyla existed in the Precambrian age, though in the Burgess Shale of British Columbia scientists identified 'eight anatomical designs that do not fit into any known animal phylum ... But this list is nowhere near complete ... The best estimates indicate that only about half the weird wonders of the Burgess Shale have been described'.[4] We don't know if the total number of Precambrian phyla throughout the world was 20, or 40, or 117. While the number is not infinite, a realistic bound is unknown, even assuming that the concept of a phylum is well defined. Furthermore, we have no idea *why* the preponderance of Burgess phyla became extinct: 'we have no evidence whatsoever – not a shred – that losers in the great decimation were systematically inferior in adaptive design to those that survived'.[5] That this assertion may be hotly disputed by some biologists only strengthens the point.

The unknown is not limited to the physical or biological worlds, but includes the realm of human creativity as well. We don't know the contents of the lost book *Porisms*, nor what Archimedes would have discovered next had he not fallen to that Roman soldier's sword. As a more modern example, 'no one knows, really knows, that is, what role nuclear weapons would play in war'.[6] We don't even know how many nuclear weapons have been manufactured or where they all are. The possible answers to these few questions are endless, and they all reside in the realm of the unknown.

The next thing to say about the unknown is that it is in no way limited by the strictures of logical consistency. For instance, the known mass of the earth cannot be both 6×10^{24} kg and *not* 6×10^{24} kg. If knowledge is to be distinguished from error or ignorance, then it cannot contain contradictions (because from a contradiction one can deduce any other statement). However, regarding the mass of an unknown planet that might exist at the other edge of our galaxy, both statements have equal status in the realm of the unknown: one or the other is true, but we don't know which. Furthermore, the unknown nourishes the imagination and is the subject of science fiction as much as it is the subject of science.

We can hold in our minds thoughts of four-sided triangles, parallel lines that intersect, and endless other seeming impossibilities, from super-girls like Pippi Longstocking[7] to life on Mars or merciful murder (some of which may actually be true, or possible, or coherent).[8]

Scientists, logicians and saints are in the business of dispelling all such incongruities, errors and contradictions. Banishing inconsistency is possible in science because (or if) there is only one coherent world. Banishing hypocrisy and deceit is possible if people follow particular moral codes. The unknown is the realm in which scientists and saints have not yet completed their tasks. For instance, we must entertain a wide range of conflicting conceptions when we do not yet know how (or whether) quantum mechanics can be reconciled with general relativity, or Pippi's strength reconciled with the limitations of physiology, or killing with murder. As Henry Adams wrote:

> Images are not arguments, rarely even lead to proof, but the mind craves them, and, of late more than ever, the keenest experimenters find twenty images better than one, especially if contradictory; since the human mind has already learned to deal in contradictions.[9]

A theory, in order to be scientific, must exclude something. A scientific theory makes statements such as 'this happens; that doesn't happen'. Karl Popper explained that a scientific theory must contain statements that are at risk of being wrong, statements that could be falsified.[10] Deborah Mayo demonstrated how science grows by discovering and recovering from error.[11]

The realm of the unknown contains contradictions (ostensible or real), such as the pair of statements: 'Nine year old girls can lift horses' and 'Muscle fiber generates tension through the action of actin and myosin cross-bridge cycling'.

Scientific theories cannot tolerate such contradictions. But it is a mistake to think that the scientific paradigm is suitable to all activities, in particular, to thinking about the unknown. Logic is a powerful tool and the axiomatic method assures the logical consistency of a theory. For instance, Leonard Savage argued that personal probability is a 'code of consistency' for choosing one's behaviour.[12] In contrast, Jim March compares the rigorous logic of mathematical theories of decision to strict religious morality. Consistency between values and actions is commendable says March, but he notes that one sometimes needs to deviate from perfect morality. While '[s]tandard notions of intelligent choice are theories of strict morality ... saints are a luxury to be encouraged only in small numbers'.[13] Logical consistency is a merit of a scientific theory. However, logical consistency does not constrain the unknown because the unknown is the reservoir of possibilities and the source of inventiveness.

Info-Gap Theory

The unknown presents serious challenges to decision-makers. Knight, in discussing entrepreneurial decision-making, distinguished between 'risk' based on known probability distributions and 'true uncertainty' for which probability distributions are not known.[14] Wald, in studying statistical decisions, also considers situations of ignorance that cannot be represented by probability distributions.[15] Lempert, Popper and Bankes and many others studying public policy, such as Beck in this volume, discuss severe or 'deep' uncertainty that reflects the unknown.[16]

Info-gap theory is a decision theory: a methodology for modelling and managing the unknown and for supporting the formulation of plans, strategies, designs and decisions under severe uncertainty.[17] In this subsection we will describe the basic concepts of info-gap theory. We will be speaking of human decision-making, and on pp. 170–4 we will refine our language and attempt to remove the anthropomorphism when we apply info-gap theory to both epistemic and ontological indeterminism.

Decision-makers are goal-oriented. The engineering designer aims to increase the payload of an aircraft, or to decrease the failure rate of a milling machine. The economic planner seeks to improve profits or to reduce unemployment. The physician seeks to prevent or cure disease. The military strategist seeks to deter the enemy or to win in armed conflict if deterrence fails.

Decision-makers do not operate in total darkness. They have data, knowledge and understanding, which we will collectively refer to as 'models'. Sometimes those models are confidently known, as in choosing rain gear for a winter trip to London: take an umbrella! At other times the models are imperfect but their uncertainties are highly structured. For instance, the farmer can choose next season's crops based on reliable statistical data of temperatures and rainfall in recent years. Sometimes the uncertainties are severe and unstructured. Predicting rainfall or storm patterns a century hence is plagued by substantial lack of understanding about the processes of global warming. Info-gap theory is useful in modelling and managing severe uncertainty.

An *info-gap* is the disparity between what one *does know*, and what one *needs to know* in order to make a reliable, responsible decision. An info-gap is part of the disparity between the known and the unknown.

An *info-gap model of uncertainty* is a mathematical device that quantifies an info-gap. An info-gap model is an unbounded family of nested sets of possible realizations of an unknown entity. The unknown entity may be tomorrow's stock value, or the number of Precambrian phyla, or the functional relation between temperature and fertility of an insect species, or an adversary's unknown preferences among possible outcomes of a future nuclear war, or the probability that a new veterinary disease will spread to humans, or the probability distribution of useful lifetimes of a new micro-robotic machine. Mathematically, the elements of the sets of an info-gap model of uncertainty are numbers, or vectors, or functions, or sets of such things. An info-gap model is not a single set, but rather an unbounded family of nested sets. The sets become more and more inclusive, reflecting a growing horizon of uncertainty. The sets grow without bounds (within the domain of their definition), reflecting the unknown bounds on possible realizations of the uncertain entity. An info-gap model represents the unknown degree of error of the decision-maker's models.

Two concepts are central in info-gap theory in supporting the decision-maker's choice of an action in attempting to achieve a goal in the face of severe uncertainty: robustness and opportuneness.

The *robustness* of a decision assesses its tolerance to the unknown for achieving an outcome requirement. A decision is highly robust if the decision-maker's goal will be achieved despite vast uncertainty about the decision-maker's models. A decision is not robust if low uncertainty jeopardizes the achievement of the goal.

We now consider *opportuneness*. Decision-makers are goal-oriented, but sometimes the goal is not a requirement (or demand, or obligation) such as 'The useful lifetime of the system will not be less than thirty years'. Sometimes the goal is an aspiration for windfall or better-than-anticipated outcomes, like 'It would be wonderful – though it's not a design requirement – if the useful lifetime of the system exceeds fifty years'. A decision is opportune if wonderful windfalls are possible – though not necessarily guaranteed – even at low levels of uncertainty. A decision is opportune if small deviations of reality from the models can facilitate large windfalls. While the robustness of a decision is its immunity against failure to achieve the goal, the *opportuneness* of a decision reflects its potential for wonderful windfall.

The decision-maker who seeks to achieve a goal, despite severe uncertainties in the models, faces two irrevocable trade-offs, one regarding robustness against uncertainty, and the other regarding opportuneness from uncertainty.

Robustness to error trades off against the quality of the goal. The decision-maker must choose an action (or design, or strategy, etc.) and can use the models to predict whether or not a contemplated action would achieve the goal. The models represent the best available data, knowledge and understanding, but they are accompanied by severe uncertainty and thus may err greatly. Hence predicted outcomes are not a reliable basis for decision. More specifically, it is not unreasonable to expect that the actual outcome may be worse that the predicted outcome. However, a goal that is somewhat less demanding or somewhat less ambitious than the predicted outcome may be achieved even if the models err a bit. As the goal becomes more modest, the contemplated decision (action) can tolerate greater error in the models and still achieve the goal. In short, the robustness (to error, uncertainty, ignorance or surprise) *increases* as the quality of the required goal *decreases*. This is the first trade-off.

Opportuneness from error trades off against the quality of the aspiration. The decision-maker's models predict the outcome of a contemplated action. The models represent the best available data, knowledge and understanding, but they are accompanied by severe uncertainty and thus may err greatly. Hence it is not unreasonable to expect that actual outcomes may be better than predicted. An outcome-aspiration that is somewhat more ambitious than the predicted outcome may be achieved (but cannot be guaranteed) if the models err a bit. As the aspiration becomes more ambitious, the contemplated decision (action) requires greater error in the models in order for achievement of the aspiration to be possible. In short, the opportuneness (from error, uncertainty, ignorance or surprise) *decreases* as the quality of the aspiration *increases*. This is the second trade-off.

It is entirely correct that assessment of the robustness or the opportuneness of a decision depends on the decision-maker's models. In fact, that's usually the point of performing the assessment. Given the decision-maker's current state of knowledge – as expressed by the models – we wish to assess the vulnerability (or opportuneness), of a contemplated decision, to error in that knowledge. The robustness is used to prioritize the available decisions given the existing models: a more robust option is preferred over a less robust option. The opportuneness is used similarly: more opportuneness is preferred over less opportuneness (the robustness and opportuneness prioritizations need not agree). From a practical point of view – and decision-makers are usually pragmatic – a choice must be made from a given state of knowledge. What the decision-maker needs to know is how robust, or how opportune, each option is, given the available models. A decision theory in which the robustness and opportuneness were *not* sensitive to the decision-maker's models would not be useful at all.

It is incorrect to assert that, because the info-gap robustness depends on the decision-maker's models, it is local rather than global. Info-gap models of uncertainty are unbounded on the domain of their definition, so they are not constrained 'locally' to the region around the decision-maker's models. This allows the robustness to be either small or large. If the robustness is large then the goals are achieved even if reality deviates enormously from the models. At the extreme (which sometimes occurs), infinite robustness implies that the goal is guaranteed, regardless of the degree of error. At the other extreme, very small robustness implies that only small deviation is tolerable: the goal will be achieved only if reality is close to the model. That the robustness of some decisions is small does not mean that the info-gap robustness is inherently local. The robustness may be small or large, depending on the decision-maker's models and requirements. Small or large robustness implies that the domain of reliability of the contemplated decision is local or not, respectively.

As we have explained, the robustness is used by decision-makers to prioritize the available options: the more robust option is preferred over the less robust option. However, the first trade-off that we discussed asserted that the robustness of an option depends on the required quality of the goal. Hence the prioritization of the options may depend on the decision-maker's goal. It can happen, and often does, that the prioritization of the options is different for different quality requirements: a very demanding requirement might be most robustly achieved with one option, while a less demanding goal may be robust-preferred with a different option. This 'reversal of preference' is very important in decision-making under uncertainty[18] and in explaining a number of anomalies of human decision-making, including the Ellsberg and Allais paradoxes and the equity premium puzzle[19] and the home bias paradox.[20] We will discuss an example on pp. 171–3.

Ontological Indeterminism

Polarized Photons: An Example of Ontological Indeterminism

Dirac discusses the interaction between polarized photons and the polarizing crystal tourmaline.[21] If a beam of polarized light impinges at an angle α to the crystal axis, then a fraction $\sin^2\alpha$ will be transmitted and will be polarized perpendicular to the crystal axis upon exiting the crystal. If a single photon impinges, polarized at an angle α to the crystal axis, then it will either be completely absorbed or completely transmitted; in the latter case it will be polarized perpendicular to the crystal axis. A fraction $\sin^2\alpha$ of such photons, impinging independently, will be transmitted. After describing these observations Dirac writes:

> Thus we may say that the photon has a probability $\sin^2\alpha$ of passing through the tourmaline and appearing on the back side polarized perpendicular to the axis and a probability $\cos^2\alpha$ of being absorbed. These values for the probabilities lead to the correct classical results for an incident beam containing a large number of photons.
>
> In this way we preserve the individuality of the photon in all cases. We are able to do this, however, only because we abandon the determinacy of the classical theory. The result of an experiment is not determined, as it would be according to classical ideas, by the conditions under the control of the experimenter. The most that can be predicted is a set of possible results, with a probability of occurrence for each.
>
> The foregoing discussion about the result of an experiment with a single obliquely polarized photon incident on a crystal of tourmaline answers all that can legitimately be asked about what happens to an obliquely polarized photon when it reaches the tourmaline. Questions about what decides whether the photon is to go through or not and how it changes its direction of polarization when it does go through cannot be investigated and should be regarded as outside the domain of science.[22]

Dirac's discussion illustrates what we are calling ontological indeterminism: 'The result of an experiment is not determined ... by the conditions under the control of the experimenter.' The lack of control is not the experimenter's deficiency, but rather nature's indeterminism. The classical conception of a law of nature is modified by quantum mechanics. When an event occurs it does so in one way and not any other (excluding multiple worlds interpretations). To this extent, some 'law' or property inherent in the substances and circumstances may be said to be acting. However, if identical substances and circumstances at different instants or locations result in non-identical outcomes, then 'natural law' as understood classically – immutable and universal – does not hold. There is some variability or indeterminism in those attributes that govern the course of events. What physicists have classically thought of as constant and fully specifiable laws are in fact to some extent indeterminate. It is not the experimenter's competence that is compromised; nature's competence to govern events is incomplete.

The Success of Science and (or Despite) Ontological Indeterminism

What can we say about the character of ontological indeterminism? If we follow Dirac (and I will), such questions are 'outside the domain of science' and are therefore speculative or philosophical. We are trying to understand how science – the search for laws of nature – works if nature has a non-nomological element.

We require concise qualitative understanding of two concepts from physics: the Lagrangian and the action integral.

The Lagrangian represents the physical properties of the system, and is often the difference between kinetic and potential energy. The physical properties of the system – whether it's a mass attached to a spring, a particle pulled by gravity, a photon in an electro-magnetic field, or whatever – are represented by the Lagrangian. The 'action integral' is the integral over time of the Lagrangian.

The equations of motion of both classical physics and quantum mechanical systems can be derived by finding a stationary point of the action integral.[23] Since the stationary point is usually a minimum, this is sometimes called the principle of least action.

Dirac does not dispute or abandon the least-action principle, and Feynman's derivation of Schrödinger's equation, based on the path-integral method, explicitly exploits the least-action principle.[24]

The problem we are facing is this. As we explained on p. 164, nature is indeterminate to some extent; it lacks a complete universal and invariant set of laws that govern physical processes; nature is non-nomological to some degree. As Dirac writes '[q]uestions about what decides whether the photon is to go through or not and how it changes its direction of polarization when it does go through cannot be investigated and should be regarded as outside the domain of science'. And yet, the Lagrangian represents the physical properties of the system. Classically this meant that the Lagrangian is obtained from the relevant laws of nature, such as stress-strain relations or gravitational attraction or Maxwell's equations. How can we reconcile ontological indeterminism and partial lack of law, with a unique and specified Lagrangian?

Classical physics dictated that nature 'finds' the unique solution, or path, that obeys the laws of nature. Feynman, in his beautiful derivation of quantum mechanics, recognizes that an infinity of different paths all contribute to the physical process. The second axiom in Feynman's paper states:

> *The paths contribute equally in magnitude [to the wave function], but the phase of the contribution is the classical action (in units of h); i.e., the time integral of the Lagrangian taken along the path.*

That is to say, the contribution $\Phi(x)$ from a given path $x(t)$ is proportional to $\exp(i/h) S[x(t)]$, where the action $S[x(t)] = \int L(\dot{x}(t); x(t))dt$ is the time integral of the classical Lagrangian $L(\dot{x}; x)$ taken along the path in question.[25]

Herein lies Feynman's reconciliation of quantum indeterminacy with the classical least-action principle. Consider two slightly different paths, x_1 and x_2, whose action integrals, $S[x_1(t)]$ and $S[x_2(t)]$, are slightly different. The phases of their contributions to the wave function will be different, so these paths will tend to cancel one another out. However, there is some path for which the classical action is stationary: for which $S[x(t)]$ does not change as $x(t)$ changes slightly. Consider a bundle of paths near this classical least-action path. These paths all have nearly the same action, and hence nearly the same phase in their contributions to the wave function. These paths will not cancel one another out, but rather re-enforce one another. Consequently, a bundle of paths near the classical path will be more likely to occur than a bundle of paths further from the classical path. Quantum indeterminacy is reconciled with classical least-action.

The classical Lagrangian is incorporated in Feynman's derivation of quantum mechanics, but it has lost its classical sovereignty and no longer determines a unique natural process in response to a unique set of initial conditions. The physicists are happy, and rightly so, because quantum mechanics has enjoyed a long history of success. But from the speculative point of view, how are we to understand the concept of a law of nature? I suggest that natural law is indeterminate; it does not take a specific functional form for any given physical system. Each path has its value of the action integral, based on the classical Lagrangian, but this does not determine the outcome of an experiment by application of the least-action principle, nor does the classical Lagrangian reflect an immutable and universal law of nature. Feynman's path-integral method, exploiting the phase-cancellation phenomenon, is needed to bridge the gap between classical determinism and quantum indeterminism.

Specific events happen specifically, at least when we observe them (quantum mechanics is circumspect about events that are not observed). When a specific event happens in one way rather than another, the classical physicists offered a Lagrangian – and the least-action principle – in explanation. This programme collapsed with the advent of quantum mechanics. Ontological indeterminism is the gap between our classical understanding of the world (the classical Lagrangian) and some hypothetical law of nature that governs any specific event. This gap is unbounded in the sense that the 'hypothetical law of nature' may have no real existence, no ontological status. The gap is also unbounded in the sense that all path-bundles contribute quantum mechanically to some (unequal) extent. The gap that I am calling ontological indeterminism results in the constriction of the domain of science to which Dirac attests. In the quantum world, no single path 'wins' universally, in a given experimental setup, and no single Lagrangian or law of nature 'rules' in the classical sense. The range of paths that participate in the physical event is unbounded. Ontological indeterminism is the gap between what we do know – the classical Lagrangian and the least-action principle – and what the classical physicist would want to know – a complete explanation of specific events. This gap in our understanding is unbounded and, if the quantum mechanicists are right, unbridgeable.

Epistemic Indeterminism

The history of humanity is, in large measure, the story of discovery and invention. Our proud self-images as homo sapiens, the Wise Man and the Tool Maker, convey the central message of deliberation, innovation and change. Despite the deliberative element, discovery and innovation entail surprise, and this has far-reaching implications.

Habermas[26] emphasizes the non-nomological nature of social science, and Nelson and Winter stress that in evolutionary economics 'things always are changing in ways that could not have been fully predicted'.[27] We learn things, which Keynes referred to as hearing the 'news', and this sometimes dramatically alters our behaviour.[28] The idea of indeterminism in human affairs was developed separately and in different ways by Shackle[29] and Popper.[30] We will refer to this Shackle–Popper indeterminism as epistemic indeterminism because it derives from the limitation of what we know.[31]

The basic idea of Shackle–Popper indeterminism is that the behaviour of intelligent learning systems displays an element of unstructured and unpredictable indeterminism. By 'intelligence' I mean: behaviour is influenced by knowledge. This is surely characteristic of humans individually and of society at large. By 'learning' I mean a process of discovery or invention: finding out today what was unknown yesterday. Finally, indeterminism arises as follows: because tomorrow's discovery is by definition unknown today, tomorrow's behaviour is not predictable today, at least not in its entirety. Given the richness of future discovery, (or its corollary, the richness of our current ignorance), the indeterminism of future behaviour is broad, deep and unstructured. The laws of human behaviour will change over time as people make discoveries and inventions. These laws cannot be known in their entirety ahead of time, because by definition discoveries cannot be predicted and the laws of behaviour depend on the discoveries to be made.

Epistemic indeterminism is the gap between what we *do know* about ourselves, our world and our future, and what we would *need to know* in order to fully grasp and control that trajectory. This gap is unbounded as long as the universe of possible future discovery and invention is open-ended and accessible. This gap is continually shifting as new knowledge emerges, but the frontier of possibilities is shifting as well.

In the first section we distinguished between ontological and epistemic indeterminism, and in this and the previous section we elaborated on these two conceptions. Nonetheless, our main claim is the structural similarity between these vastly different phenomena, unified by the concept of info-gap robust-satisficing. We now begin to explore the underlying commonality between epistemic and ontological indeterminism.

Optimizing, Satisficing and Indeterminism

Optimizing

Optimization – finding stationary values of an objective function – is a fundamental concept in physics. The variational principles of mechanics are optimization problems, equilibrium in thermodynamics involves minimizing an energy function, and so on. The 'optimization paradigm' asserts that laws describe the behaviour of a system and that these laws can be derived by optimizing a physically meaningful objective function. The properties of a system are embodied in its objective function. Examples of objective functions in physics are the action integral or the Gibbs free energy.

The optimization paradigm is prominent in the biological and social sciences as well, where the optimization of substantive or useful outcomes is a normative explanatory tool. For example, a large body of biological literature seeks to explain the foraging behaviour of animals as guided by maximizing caloric intake.[32] Fitness or some measure of survival is often optimized in theories of evolutionary competition. In social science, mathematical economists derive equations of motion of an economy from 'first order conditions' that specify optimality of economic utility functions.[33]

Satisficing and the Limits of Optimization

Optimization requires that the objective function be determinate and accessible in some sense. This does not mean that the agents involved (projectiles, pigeons or pawn brokers, for instance) need conscious knowledge of the objective function. Nonetheless, that function must be stable and discoverable – perhaps implicitly – in some relevant sense. Pigeons need not be conscious of their caloric intake, but measures or expressions of that intake must be accessible to them in some way, for instance by the evolutionary selection process against pigeons whose caloric intake is too low.

The need for stability and accessibility of the objective function limits both the explanatory power and the practical utility of the optimization paradigm in social science. Simon studied what he called the 'bounded rationality' of animal, human and organizational decision-makers.[34] Bounded rationality is the behavioural consequence of limited access to information and understanding, and the limited ability to process that information. As a consequence, Simon claimed, organisms 'satisfice'. Etymologically, 'to satisfice' is a variant on 'to satisfy', but 'satisfice' has come to have a tighter technical meaning in economics, psychology and decision theory. The Oxford English Dictionary defines 'satisfice' in this technical sense to mean 'to decide on and pursue a course of action that will satisfy the minimum requirements necessary to achieve a particular goal.'[35]

The prevalence of the outcome-optimization paradigm in the study of human affairs is due at least in part to the fact that more of a good thing is, by definition, better than less. Of course, things are not so simple. Optimizers are well aware of diminishing marginal utility, of substitution between goods, of budget constraints, and of absurd extremes (chocolate is good; too much chocolate is a belly ache). Theories of optimization account for these subtleties by modifying the objective function and by imposing constraints on the domain of solution. The limitations of the optimization paradigm in the social sciences derive from epistemic indeterminism, not from the subtle and multi-variate nature of human preferences. Optimization is of limited utility or feasibility if either information or information-processing capabilities are too scarce.

The attractiveness of the optimization paradigm in physics derives from various historical and intellectual factors that we cannot explore here. Very briefly, though, one motivation is certainly the beauty and simplicity of deriving all laws of nature from a single concept: the least-action principle. Another speculative motivation may be related to the search for certainty in a confusing world. If God is dead, (or at least dying as in the early stages of modernity), then the optimal perfection of natural law is not a bad surrogate.

The limitation of optimization in physics is subtler than in human affairs. The gist (and the beauty) of Feynman's derivation of the Schrödinger equation, as discussed on pp. 165–7, is that the least-action principle is preserved, though its impact is muted. The path whose action integral is stationary (this is the unique classical path) is the quantum mechanically dominant (most likely) path. Other (classically forbidden) paths contribute quantum mechanically with diminishing strength as their action integrals become less stationary. The classical optimization that leads to a unique state of nature, is replaced by an infinity of sub-optimal solutions that all contribute to some degree as possible physical realizations. The discovery of ontological indeterminism led to constriction of the domain of science, and dilution of the concepts of causality and optimality, as discussed on pp. 164–5.

The success of satisficing – as an alternative to optimizing – in understanding the physical, biological and human worlds, derives from the nature of indeterminism that underlies the course of events in all these domains. The info-gap conception of uncertainty, and the resulting quantification of robustness, provide a framework for characterizing this indeterminism and the responses to it as we will discuss in the next section.

Info-Gap Robust-Satisficing and the Response to Indeterminism

Knight's discussion of economic behaviour under severe uncertainty distinguishes between 'risk' based on known probability distributions and 'true uncertainty' for which probability distributions are not known.[36] The info-gap conception of uncertainty, described on pp. 158–60, is very Knightian. Info-gap robust-satisficing and opportune windfalling, discussed on pp. 161–4, are operational responses to Knightian uncertainty. Info-gap theory[37] has been applied to decision problems in many fields, including various areas of engineering,[38] biological conservation,[39] economics,[40] medicine,[41] homeland security,[42] public policy[43] and more (see info-gap.com).

We are now in a position to explain how the concepts of satisficing and robustness to info-gap uncertainty provide a unified framework for understanding both epistemic and ontological indeterminism. This requires us to transcend the anthropomorphic and teleological language that we have used and that often characterizes the discussion of theories of human decision-making. Human language, as distinct from mathematics, is embedded in human experience, so this transcendence will be only partial. Nonetheless, the following subsection introduces the basic definitions that we will use in attempting to ameliorate the anthro-centrism of our discussion. The two subsequent subsections discuss epistemic and ontological indeterminism.

Terminology for Info-Gap Uncertainty and Robust-Satisficing

We will talk about '*a world*' as an entire sequence of events, both past and future. There are many possible worlds in the sense that many sequences of events are possible.

We will talk about '*decisions*' without necessarily entailing consciousness or volition. A decision has been made, in the sense we intend, if the course of events in the world goes one way rather than another. A person's decision may result from that person's volitional choices. But we can also talk about the decisions made by an organization, or by a squirrel or a squid, or by nature as a whole.

We will talk about '*satisfying a requirement*', without necessarily implying either volition or teleology. A requirement is satisfied when a condition is met. A person's condition is met, according to the song 'All You Need is Love',[44] by achieving love. A squirrel satisfies a caloric requirement by eating acorns, and nature needs action integrals that are stationary or nearly so.

A decision is '*robustly satisficing*' if the requirement is satisfied over a wide range of possible worlds. A person's decision to visit a library, (rather than a bar perhaps), is robust if love is achieved for any of a large set of other visitors. A squirrel's decision to eat acorns under this tree (rather than under another tree) is robust if the caloric requirement is satisfied regardless of weather, or predators, or competing squirrels. Nature's decision is robust if its action integral is nearly stationary over a wide range of possible Lagrangians.

Robust-Satisficing and the Response to Epistemic Indeterminism

In human affairs, robust decisions are motivated by epistemic indeterminism. We don't know everything that we need to know, or we don't have the capability to process all the available information, so we want to satisfy our requirements even if we err greatly about the world. Bounded rationality motivates robust-satisficing as a decision strategy.

The motivation for robust-satisficing can also be understood in the context of competition under uncertainty. In human affairs, competition among individuals, organizations, societies and civilizations sometimes comes with high stakes, even survival or demise. An entity whose critical requirements are satisfied despite huge surprises will tend to survive at the expense of competing entities that are more vulnerable to surprise. The same can be said about evolutionary competition of biological species.[45] The squirrel that is evolutionarily programmed to make the right decisions about foraging, resting and mating will tend to survive instead of other squirrels. In many situations, info-gap robust-satisficing is a better bet than any other strategy for satisfying one's requirements.[46]

Let's briefly consider a simplified example: public policy for managing an invasive biological species. The Light Brown Apple Moth (LBAM) is native to

Australia. It was detected in California in 2007 though some entomologists claimed that it had been present and widespread (though undetected) for a long time. In other words, there was expert dispute over whether the LBAM was in fact an invasive species or not. Furthermore, there was very substantial dispute about the economic damage that could result from the moth, either from direct crop loss or due to inter-jurisdictional trade restrictions on California crops potentially carrying the LBAM. There was also dispute over the efficacy of various means of control or eradication of the LBAM, as well as over possible adverse effects of these interventions on public health. In short, the LBAM created major controversy in both the scientific community and the general public.[47]

In our simplified example we will describe how info-gap robust-satisficing would be used to choose between various possible interventions (where doing nothing is one possibility). There are diverse risk factors that are poorly understood and whose adverse impact, in monetary terms, are highly uncertain. The analyst begins by asking what is the largest monetary loss that is tolerable (recognizing that this too may be uncertain). The outcome requirement is that the loss must not exceed this critical value. The analyst then identifies the data, knowledge and understanding – the models – that are relevant to the problem. (The models may contain probabilistic elements.) The models are used to predict the outcome of each intervention. Any intervention whose predicted outcome does not satisfy the outcome requirement is excluded. The vast uncertainties of the models are non-probabilistically represented using unbounded info-gap models of uncertainty. The robustness to uncertainty of each remaining intervention is evaluated. That is, the analyst determines, for each remaining intervention, how large an error in the models can be tolerated without jeopardizing the critical outcome requirement. The analyst adopts the intervention whose robustness is greatest for satisfying the requirement.

Three general characteristics of info-gap robust-satisficing are illustrated by this example.

First, the robustness of each possible intervention decreases as the outcome requirement (maximum tolerable monetary loss) becomes more demanding. This is the robustness trade-off discussed on p. 162.

Second, it could happen that all interventions that are examined have zero or very low robustness at the specified outcome requirement (maximum tolerable loss). This could induce the analyst to seek additional options, or to improve the models or to reduce their uncertainties. Alternatively, the outcome requirement might be declared unrealistic. The robustness trade-off might then be used in revising the requirement.

Third, different interventions might be most robust (and hence preferred) for different values of the maximum tolerable loss. This preference reversal among the options is particularly common when choosing between a widely used 'state of the art' option, and a new and innovative alternative. The innovation is predicted to provide a better outcome (lower loss) than the state of the art; that's what makes it attractive. However, the innovation is usually less well known and more uncertain than the state of the art which has been widely used. The innovation may have positive robustness for a very demanding requirement, at which the state of the art may even have zero robustness. However, the greater uncertainty of the innovation causes its robustness to fall below the robustness of the state of the art at less demanding requirements. The 'innovation dilemma' facing the analyst is to choose between a more promising but more uncertain option (the innovation) and a less promising but less uncertain option (the state of the art). The innovation dilemma is resolved by choosing the more robust alternative for the required outcome.[48]

Robust-Satisficing and the Response to Ontological Indeterminism

From the perspective of classical physics, the impact of ontological indeterminism can be summarized by noting that the unique classical Lagrangian no longer determines a unique response (path) to a specified set of conditions. Feynman used the classical Lagrangian in his derivation of Schröderinger's equation, though he did not use it in the way that classical physicists used it. Classically, the Lagrangian determined a unique solution, while Feynman (and quantum mechanics) allows all paths to contribute, each according to the degree of stationarity of its action integral.

The robust-satisficing response to ontological indeterminism begins by hypothesizing that the Lagrangian is indeterminate; it does not take a specific functional form for any given physical system. Consequently, a least-action principle is no longer immediately applicable: there is no specific Lagrangian with respect to which the action can be stationary. Feynman's adoption of the classical Lagrangian in a quantum setting is to be viewed as an ad hoc synthesis.

In the context of robust-satisficing, the requirement that nature must satisfy is that the action integral must be nearly stationary. What Feynman proposed is that, quantum mechanically, path-bundles are favoured whose action integrals are relatively stationary with respect to path variation; path-bundles with highly non-stationary action integrals are rare. We interpret this as a response to the ontological indeterminism of the Lagrangian: nature robustly satisfices the stationarity of the action integral.

Feynman's derivation of the Schrödinger equation remains unchanged, except that now it is the magnitude (or, more generically, stationarity) of the robustness, not stationarity of the action integral, that determines the quantum mechanical weight of a wave function. We hypothesize that the classical action depends on the path only through the robustness. Consequently, we obtain the same wave functions as in Feynman's method, but for a different reason. This puts ontological indeterminism in the same conceptual framework as epistemic indeterminism: maximize the robustness to indeterminism while satisfying a requirement. This interpretation of quantum mechanics also sheds light on issues of locality, realism and completeness.[49]

Optimization is still central because stationarity of the robustness determines the equations of motion. But the robustness is not itself a property of the system, unlike the Lagrangian in both classical and quantum mechanical physics. The Lagrangian, that embodies specific properties of the system, is indeterminate, and only the robustness to this property-indeterminism is optimized in deriving equations of motion.

Conclusion

It is amazing that we can unravel the Rosetta Stone, cosmological evolution and the neurophysiology of primate brachiation. We discern patterns, mechanisms, causes, effects and other relations. Over time, much more will be discovered and understood. Some of our current ignorance, like the location of the Ten Lost Tribes, has been resolved indirectly by understanding the historical processes of conquest and assimilation. Other quests, like a cure for HIV/AIDS, still demand great effort and will be achieved, if at all, by ways unknown today. And some quests, like unifying our understanding of the physical world, seem to evade our best efforts.

It is no less amazing that the formless void of what we don't know – even what we cannot know scientifically – seems to have a handle that we can grasp conceptually. As a start, the indeterminism of the physical, biological and human worlds is real. Ontologically, the world and its laws are not fixed and finalized. Epistemically, our ability to know is limited by the fact that tomorrow's new knowledge cannot be known today, from which the Shackle–Popper indeterminism establishes strong implications for explanation and prediction in human affairs. Second, both epistemic and ontological indeterminism are (to the best of our knowledge) unbounded: the epistemic gap between knowledge and possible knowledge, and the ontological gap between what has occurred and what will occur, has no limit. The potential for surprise is unbounded.

That these indeterminisms are manageable is evident from the simple fact that we, and nature, manage, either literally or metaphorically. Stars shine and people ramble along despite the limitation of classical causality and of human comprehension. *How* we manage is the amazing thing. Any attempt to explain this management is subject to the inevitable fallibility of all explanation[50] (and subject to the inevitable disputes among explainers). The suggestion in this chapter is that an underlying common theme, in response to both epistemic and ontological indeterminism, is to robustly satisfice the achievement of specific requirements. The concept of robust-satisficing reflects what we can – and cannot – say about indeterminism. Nature does not have *one* Lagrangian that determines *the* response, nor do people know future discoveries and inventions. But we know that we don't know the future, and nature allows multiple paths as though each were, for the moment, *the* solution. At the risk (again) of anthropomorphism, we suggest that robust-satisficing 'embraces' indeterminism: robustness is the way to 'get along' in the face of inevitable surprise.

8 LEARNING FROM DATA: THE ROLE OF ERROR IN STATISTICAL MODELLING AND INFERENCE

Aris Spanos[1]

Introduction

'How do we learn from data about phenomena of interest?' is the key question underlying empirical modelling in any discipline. The discussion that follows focuses primarily on economics, but similar problems and issues arise in all applied fields. 'Learning from data' has a long and ambivalent history in economics that can be traced back to the seventeenth century. Despite the fact that statistics and economics share common roots going back to 'political arithmetic' in the mid-seventeenth century, the relationship between them has been one of mutual distrust and uneasy allegiance.[2] They developed into separate disciplines after the demise of political arithmetic in the early eighteenth century, due mainly to:

(a) the inability of political arithmeticians to distinguish between real regularities and artefacts in observed data, combined with

(b) the irresistible proclivity to misuse data analysis in an attempt to make a case for one's favourite policies.

These excesses led political arithmeticians to exorbitant speculations and unfounded claims based on misusing the information in the data to such an extent that it eventually discredited these data-based methods beyond salvation.[3]

Political economy (economics) was separated from political arithmetic, and established itself as the first social science at the end of the eighteenth century,[4] partly as a result of emphasizing the role of theory, causes and explanations, and attributing a subordinate role to the data. Political economists made their case by contrasting their primarily deductive methods to the discredited data-based (inductive) methods utilized by political arithmeticians.

In the early nineteenth century, statistics emerged as a 'cleansed' version of political arithmetic by shaking off the discredited data-based inferential component and reassuring the critics by focusing exclusively on the collection and tabulation of data. The founding document of the Statistical Society of London in 1834 separated statistics from political economy as follows:

> The Science of Statistics differs from Political Economy because although it has the same end in view, it does not discuss causes, nor reason upon probable effects; it seeks only to collect, arrange, and compare, that class of facts which alone can form the basis of correct conclusions with respect to social and political government.[5]

The idea was that statistics would prepare the data in a variety of forms for other disciplines.

Mill[6] was the first philosopher to articulate a methodology of economics that emphasized deductively derived propositions and attributed a subordinate role to data, in an attempt to accommodate Ricardo's clear deductive perspective.[7] He argued that causal mechanisms underlying economic phenomena are too complicated – they involve too many contributing factors – to be disentangled using the available observational data. This is in contrast to physical phenomena whose underlying causal mechanisms are not as complicated – they involve only a few dominating factors – and the use of experimental data can help to untangle them by 'controlling' the 'disturbing' factors. Hence, economic theories could not establish precise enough implications of theories. At best, one can demonstrate general tendencies whose validity can be assessed using observational data. These tendencies are framed in terms of the primary causal contributing factors with the rest of the numerous (possible) disturbing factors relegated to *ceteris paribus* clauses whose appropriateness cannot, in general, be assessed using observational data. This means that any empirical evidence that contradict the implications of a theory can always be explained away as due to counteracting disturbing factors. As a result, Mill[8] attributed to statistics and data analysis the auxiliary role of investigating the *ceteris paribus* clauses in order to shed light on the disturbing factors unaccounted for by the theory which prevent the establishment of the tendencies predicted by the theory in question.[9]

More recently, the view that economic phenomena are too complicated and heterogeneous to be amenable to traditional statistical methods using observational data was reiterated by Keynes[10] in his criticisms of Tinbergen's work on the business cycle.[11] Among other less important issues and a number confusions, Keynes raised several foundational problems associated with the use of linear regression in learning from data, including:[12]

i.   the need to account for all the relevant contributing factors (including non-measurable ones), at the outset,

ii.   the conflict between observational data and the *ceteris paribus* clauses invoked by economic theories,

iii. the spatial and temporal heterogeneity of economic phenomena stemming from complex interactions among numerous economic agents,
iv. the validity of the assumed functional forms of economic relations, and
v. the ad hoc specification of the lags and trends in economic relations.

This chapter argues that both standpoints articulated by Mill and Keynes stem from an inadequate understanding of the nature and applicability of statistical induction as it relates to learning from data about phenomena of interest.

Keynes's criticism reminds one of an earlier charge by Robbins[13] – expressing a widely held view among economists – who dismissed the use of statistics for theory appraisal in economics. Their argument was that

vi. statistical techniques are only applicable to data which can be considered as 'random samples' from a particular population, i.e. Independent and Identically Distributed (IID) random variables.[14] In light of that, statistical analysis of economic data had no role to play in theory assessment.

The primary objective of this chapter is to argue that the claims (i)–(vi) are misinformed.

First, some of these claims, namely (iii) and (vi), largely ignore two important developments in the 1930s:

(a) the new frequentist statistics pioneered by Fisher, Neyman and Pearson, and

(b) the theory of stochastic processes that greatly extended the intended scope of statistical modelling to include data that exhibit both heterogeneity and dependence.

Second, the confusion between two different dimensions of empirical modelling: the *statistical* and the *substantive adequacy* pervades most of the claims (i)–(vi). Statistical adequacy has to do with whether a model accounts for the *statistical information* in the data, but substantive adequacy with whether the model sheds adequate light on (explains, describes, predicts) the phenomenon of interest. In what follows the notion of *substantive information* is used to denote any form of non data-based information, that includes theoretical and experimental information. That is, issues (i)–(ii) pertain to the substantive adequacy of the estimated model, but issues (iii)–(vi) pertain to its statistical adequacy.

The next section provides a summary of the frequentist approach to inference with a view to bring out its applicability to non-IID data, as well as shed light on its objectives and underlying reasoning in learning from data. On pp. 187–90, the Pre-Eminence of Theory (PET) perspective that has dominated empirical modelling in economics since Ricardo[15] is discussed, in an attempt to highlight its key weaknesses: (a) conflating the statistical with the substantive premises of inference, and (b) neglecting the validation of the latter that would secure the reliability of inference, in an attempt to motivate the need for a broader methodological framework that allows for the data to play a more prevalent role. The error statistical framework, initially proposed by Mayo,[16] is presented in the section on Error Statistics: A Modelling Framework (pp. 190–206) as a refinement/extension of the Fisher–Neyman–Pearson approach that helps to delineate the above issues and address some of the foundational problems bedevilling frequentist statistical inference since the 1930s.

The Frequentist Approach to Statistical Inference

In the early 1920s R. A. Fisher recast statistical inference, by breaking away from the then dominating descriptive statistics paradigm as modified/extended by Karl Pearson.[17]

The Fisher–Neyman–Pearson (F–N–P) Paradigm

Fisher[18] pioneered modern frequentist statistics as a model-based approach to statistical induction anchored on the notion of a statistical model:

$$\mathcal{M}_\theta(\mathbf{z}) = \{f(\mathbf{z}; \theta), \ \theta \in \Theta\}, \ \mathbf{z} \in \mathbb{R}^n_Z, \ \Theta \subset \mathbb{R}^m, \ m < n, \tag{1}$$

where $f(\mathbf{z}; \theta)$ is the joint distribution of the sample $\mathbf{Z} := (Z_1, ..., Z_n)$. Fisher's most enduring contribution is his devising a general way to 'operationalize' errors by embedding the material experiment into a statistical model, and taming errors of inference by defining frequentist error probabilities in its context. These error probabilities are (a) deductively derived from the statistical model, and (b) provide a measure of the 'trustworthiness' of the inference procedure: how often a certain method will give rise to reliable inferences. The mathematical apparatus of frequentist statistical inference was largely in place by the late 1930s. Fisher[19] largely created the current theory of 'optimal' point estimation and formalized significance testing based on the p-value reasoning. Neyman and Pearson[20] proposed an 'optimal' theory for hypothesis testing, by modifying/extending Fisher's significance testing. Neyman[21] proposed an 'optimal' theory for interval estimation analogous to N–P testing.

The Fisher–Neyman–Pearson (F–N–P) model-based frequentist inference, relying on finite sampling distributions amounted to a major break from the early twentieth-century descriptive statistics paradigm, relying on vague allusions to large samples and unobtrusive priors, but the change of paradigms was neither apparent nor without its polemists. Indeed, several statistics textbooks that were influential in economics, like Bowley,[22] Mills[23] and Allen,[24] missed this change of paradigms altogether. In particular, the overwhelming majority of their chapters revolve around descriptive techniques based on a vague frequentist interpretation of probability, but without a clear probabilistic foundation, akin to the Karl

Pearson and not the F–N–P perspective. For instance, correlation and regression are viewed from a curve-fitting perspective instead of a well-defined statistical model (linear regression) with clearly specified probabilistic assumptions.

Broadening the Scope of Statistical Modelling

Specifying statistical models in terms the joint distributions $f(z; \theta)$ of a stochastic process $\{Z_t, \ t \in \mathbb{N}\}$ enables one to account for different forms of dependence and heterogeneity in data Z_0 by viewing the latter as a 'typical' realization of the stochastic process $\{Z_t, \ t \in \mathbb{N}\}$ underlying $\mathcal{M}_\theta(z)$.[25] Whether economic data exhibit sufficient *invariance* over $t \in \mathbb{N}$ to be amenable to statistical modelling and inference is an empirical issue that pertains to the validity of the statistical premises $\mathcal{M}_\theta(z)$. That is, the only restriction on the heterogeneity/dependence in the data in question is that they exhibit a sufficient *t-invariance* that encapsulates the *recurring features* of the phenomenon being modelled in the form of constant parameters θ.

The theory of stochastic processes, developed in the 1930s and 1940s, has extended considerably the intended scope of statistical modelling, by introducing various departures from IID, including Markov dependence, martingale dependence, mixing conditions, as well as several forms of heterogeneity restrictions. This broadening gave rise to numerous operational statistical models beyond IID processes. This is reflected by the extension of limit theorems (Law of Large Numbers (LLN) and Central Limit (CLT)) in the form of sufficient probabilistic restrictions (distribution, dependence and heterogeneity) on $\{Z_t, \ t \in \mathbb{N}\}$ under which 'learning from data' is potentially possible. The earlier limit theorems holding when $\{Z_t, \ t \in \mathbb{N}\}$ is IID, have been greatly extended to hold for more general stochastic processes including a Martingale Difference (MD) process, which covers numerous forms of non-IID assumptions.[26] The generality of the probabilistic assumptions for a process to obey the key limit theorems like the LLN and the CLT[27] is very important because limit theorems demarcate the scope of statistical modelling by defining conditions under which 'learning from data' is potentially possible as $n \to \infty$.

Frequentist Estimation

The key features of the F–N–P frequentist approach to inference are the following.
   (a) The interpretation of probability is grounded on the Strong Law of Large Numbers (SLLN) providing a sound link between relative frequencies and mathematical probability under certain restrictions on the probabilistic structure of the process $\{Z_t, \ t \in \mathbb{N}\}$ underlying data Z_0.[28]
   (b) The chance regularities exhibited by data Z_0 constitute the *only relevant statistical information* for the specification of the probabilistic structure

of the generic stochastic process $\{\mathbf{Z}_t, t \in \mathbb{N}\}$ underlying the data. The statistical model $\mathcal{M}_\theta(\mathbf{z})$ parameterizes $\{\mathbf{Z}_t, t \in \mathbb{N}\}$ with a view to pose the substantive questions of interest. *Substantive information* comes in the form of restrictions on statistical parameters that needs to be tested before imposed.

(c) The primary aim of the frequentist approach is to *learn from data* about the 'true' statistical data-generating mechanism $\mathcal{M}^*(\mathbf{z}) = \{f(\mathbf{z}; \theta^*)\}$, $\mathbf{z} \in \mathbb{R}^n_Z$.

In frequentist estimation learning from data using an estimator $\widehat{\theta}$ of $\theta \in \Theta$ is evaluated in terms of the sampling distribution $f_n(\widehat{\theta}; \theta^*)$ where θ^* denotes the true value of θ. The finite sample properties (any $n > 1$) are defined directly in terms of this distribution, and they include unbiasedness, full efficiency and sufficiency.

Unbiasedness. An *estimator* $\widehat{\theta}$ of q is said to be *unbiased* if: $E(\widehat{\theta}) = \theta^*$ where θ^* denotes the true value of θ in Θ, whatever that happens to be.

Table 8.1: The simple Bernoulli model

Statistical GM:	$X_t = \theta + u_t$, $t \in \mathbb{N}$,
(1) Bernoulli:	$X_t \sim \text{Ber}(.,.)$,
(2) Constant mean:	$E(X_t) = \theta$, $0 \le \theta \le 1$, for all $t \in \mathbb{N}$,
(3) Constant variance:	$Var(X_t) = \theta(1-\theta)$, for all $t \in \mathbb{N}$,
(4) Independence:	$\{X_t, t \in \mathbb{N}\}$ – independent process.

Example. In the case of the simple Bernoulli model (Table 8.1), the Maximum Likelihood Estimator (MLE) $\widehat{\theta}_n(\mathbf{X}) := \overline{X}_n = \frac{1}{n} \sum_{t=1}^n X_t$ is an unbiased estimator of θ. Evaluating its sampling distribution under the TSN ($\theta = \theta^*$) one can deduce that:

$$\widehat{\theta}_n(\mathbf{X}) \overset{\theta=\theta^*}{\backsim} \text{Bin}(\theta^*, \tfrac{\theta^*(1-\theta^*)}{n}). \tag{2}$$

Hence, factual reasoning establishes that $\widehat{\theta}_n(\mathbf{X})$ has a mean equal to the true θ^*, whatever that value happens to be. The same sampling distribution can be used to show that $\widehat{\theta}_n(\mathbf{X})$ is also fully efficient, as well as a sufficient estimator of θ.[29] Taken together these properties ensure that $f(\widehat{\theta}(x); \theta^*)$ is not only located at θ^*, but its dispersion around that value is as small as possible.

The asymptotic properties of an estimator $\widehat{\theta}$ of θ are defined in terms of the asymptotic sampling distribution $f_\infty(\widehat{\theta}; \theta^*)$ aiming to approximate $f_n(\widehat{\theta}; \theta^*)$ as $n \to \infty$, which in the above case takes the form:

$$\widehat{\theta}_n(\mathbf{X}) \overset{\theta=\theta^*}{\underset{\alpha}{\backsim}} N(\theta^*, \tfrac{\theta^*(1-\theta^*)}{n}). \tag{3}$$

Of particular interest is the asymptotic property of consistency.

Consistency (*strong*). $\widehat{\theta}_n$ zeroes-in on θ^* as $n \to \infty$, denoted by $\widehat{\theta}_n \overset{a.s.}{\to} \theta^*$:

$$\mathbb{P}(\lim_{n\to\infty} \widehat{\theta}_n(\mathbf{X})=\theta^*)=1.$$

This property is illustrated in Figures 8.1 and 8.2, where for a large enough n the estimate $\widehat{\theta}_n(x_0)$ 'points out' the true θ with probability one!

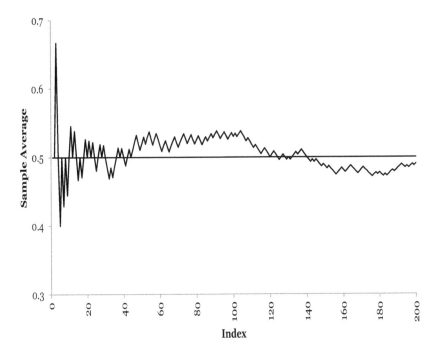

Figure 8.1: t-plot of $\overline{x}_n$ for a Bernoulli IID realization with $n = 200$.

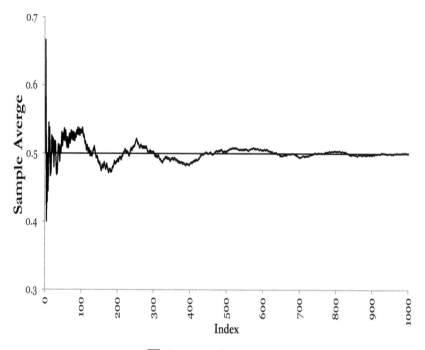

Figure 8.2: t-plot of $\overline{x}_n$ for a Bernoulli IID realization with $n = 1000$.

Consistency (weak or strong) is an extension of the Law of Large Numbers (LLN), and a consistent estimator of θ indicates potential (as $n \to \infty$) learning from data about the true but unknown parameter θ^*. Its importance stems from the fact that consistency is a minimal property of an estimator, necessary but not sufficient.

In this sense consistency demarcates the limits of statistical modelling in the sense that it specifies necessary and/or sufficient restrictions on the probabilistic structure of a stochastic process $\{X_t,\ t\in\mathbb{N}\}$ under which a potentially *operational statistical model* can be specified. However, to render the potential learning designated by consistency into actual learning, one needs to supplement consistency with certain finite sample properties like full efficiency and sufficiency to ensure that learning can take place with the particular data $\mathbf{x}_0 := (x_1, x_2, ..., x_n)$ of sample size n.

Empirical example. Let the simple Bernoulli model (Table 8.1) be selected as the appropriate $\mathcal{M}_\theta(\mathbf{x})$ for data $\mathbf{x}_0$, $n = 9869$, boys $= 5152$, girls $= 4717$. Assuming the model is valid, how do we learn from data about $\mathcal{M}^*(\mathbf{x})$, $\mathbf{x}\in\mathbb{R}_X^n$? By using the data $\mathbf{x}_0$ in conjunction with effective inference procedures to reduce the infinite parameter space $\Theta := [0, 1]$ to a smaller subset, ideally down to the point $\theta = \theta^*$, i.e. use data $\mathbf{x}_0$ as a guide to zero-in on the true $\mathcal{M}^*(\mathbf{x})$, $\mathbf{x}\in\mathbb{R}_X^n$.

Despite the fact that the sample size is very large, $n = 9869$, one should *not* infer that this estimate is 'very close' to θ^*, since in inductive inference one needs to calibrate the uncertainty associated with 'how close'.

It turns out that point estimation is often considered inadequate for the purposes of scientific inquiry because a 'good' point estimator $\widehat{\theta}_n(\mathbf{X}):=\overline{X}_n$, by itself, does not provide any measure of the reliability and precision associated with the estimate $\widehat{\theta}_n(\mathbf{x}_0)$. This is the reason why $\widehat{\theta}_n(\mathbf{x}_0)$ is often accompanied by its standard error.

Interval estimation can be used to rectify this weakness of point estimation by providing the relevant error probabilities associated with inferences pertaining to 'covering' the true value θ^* of θ. The above example provides the relevant error probabilities associated with the $(1-a)$ Confidence Interval (CI):

$$\mathbb{P}\left(\overline{X}_n - c_{\frac{a}{2}} s_{\overline{x}} \le \theta \le \overline{X}_n + c_{\frac{a}{2}} s_{\overline{x}}\right) = 1 - \alpha, \tag{4}$$

where $s_{\overline{x}}^2 = \left(\overline{X}_n(1-\overline{X}_n)\right)/n$. The inferential claim is that this random interval will cover (overlay) θ^*, *whatever that happens to be*, with probability $(1-a)$.

For the above numerical example, $1.96 s_{\overline{x}} = .009855$, thus the observed .95 CI is:
$$[.51215, .53186]. \tag{5}$$

Does this imply that the true θ^* lies within those two bounds with probability .95? Not necessarily! The observed CI (5) either includes or excludes θ^*, but we do not know. This weakness can be alleviated using hypothesis testing.

Frequentist Testing

Consider the simple normal model (Table 8.2), and let the hypotheses of interest be:

$$H_0: \mu = \mu_0 \quad \text{vs} \quad H_1: \mu > \mu_0. \tag{6}$$

The test $T_\alpha := \{d(\mathbf{X}), \mathbf{C}_1(\alpha; \mu_0)\}$, where for $\overline{X}_n = \frac{1}{n}\sum_{i=1}^{n} X_i$, and $s^2 = \frac{1}{n-1}\sum_{i=1}^{n}(X_i - \overline{X}_n)^2$:

$$d(\mathbf{X}) = \frac{\sqrt{n}(\overline{X}_n - \mu_0)}{s} \overset{H_0}{\backsim} St(n-1), \quad \mathbf{C}_1(\alpha) := \{\mathbf{x} : d(\mathbf{x}) > c_\alpha\}, \tag{7}$$

where $St(n-1)$ denotes the Student's t distribution with $(n-1)$ degrees of freedom, is Uniformly Most Powerful (UMP).[30]

Table 8.2: The simple normal model

Statistical GM:	$X_t = \mu + \mu_t,\ t \in \mathbb{N} := (1, 2, ..., n, ...)$,
(1) Normal:	$X_t \sim \mathsf{N}(.,.)$,
(2) Constant mean:	$E(X_t) = \mu$, for all $t \in \mathbb{N}$,
(3) Constant variance:	$Var(X_t) = \sigma^2$, for all $t \in \mathbb{N}$,
(4) Independence:	$\{X_t,\ t \in \mathbb{N}\}$ – independent process.

The relevant significance level and *p-value* are defined by:

$$\mathbb{P}(d(\mathbf{X}) > c_\alpha; H_0) = \alpha, \quad \mathbb{P}(d(\mathbf{X}) > d(\mathbf{x}_0); H_0) = p(\mathbf{x}_0). \tag{8}$$

The *power* of this test is defined by:

$$\pi(\mu_1) = \mathbb{P}(d(\mathbf{X}) > c_\alpha; \mu_1) \text{ for } \mu_1 = \mu_0 + \gamma_1,\ \gamma_1 \geq 0, \tag{9}$$

stemming from the sampling distribution:

$$d(\mathbf{X}) = \frac{\sqrt{n}(\overline{X}_n - \mu_0)}{s} \overset{\mu=\mu_1}{\backsim} \mathsf{St}(n-1; \delta), \quad \delta = \frac{\sqrt{n}(\mu_1 - \mu_0)}{\sigma},\ \mu_1 > \mu_0. \tag{10}$$

UMP means that this test has the highest capacity to detect any discrepancy of the form $\mu_1 > \mu_0$ than any other α–significance level N–P test.

At this stage it is important to bring out two differences between factual and hypothetical reasoning underlying (8)–(9).

First, practitioners often confuse the following two distinct sampling distributions:

$$\text{(a) } d(\mathbf{X}) = \frac{\sqrt{n}(\overline{X}_n - \mu_0)}{s} \overset{\mu=\mu_0}{\backsim} \mathsf{St}(n-1),\ \text{(b) } d(\mathbf{X}; \mu) = \frac{\sqrt{n}(\overline{X}_n - \mu^*)}{s} \overset{\mu=\mu^*}{\backsim} \mathsf{St}(n-1), \tag{11}$$

where $d(\mathbf{X})$ is a test statistic evaluated under the null, and $d(\mathbf{X}; \mu)$ is a pivot evaluated under the TSN, overlooking the fact that μ_0 in (a) is a predesignated value of μ, but μ^* in (b) denotes the true (but unknown) value of μ.

Second, in factual reasoning there is only *one* scenario, but in hypothetical reasoning there is an *infinity* of them. This has two important implications:

Testing poses sharper questions and often elicits more precise answers.

The error probabilities associated with hypothetical reasoning are also definable *post-data*, using (10). This explains why post-data CIs have degenerate coverage error; the TSN scenario has played out and the relevant error probability stemming from (11) (b) is either one or zero.

Statistical Misspecification and Unreliable Inference

A strong case can be made that the single most crucial contributor to the accumulation of untrustworthy evidence in applied econometrics is statistical misspecification.[31] What goes wrong when $\mathcal{M}_\theta(\mathbf{x})$ is statistically misspecified?

The assumed likelihood function is invalid $L(\theta; \mathbf{x}_0)$ and that gives rise to erroneous inferences. For *Bayesian* inference, it induces an erroneous posterior $\pi(\theta|\mathbf{x}_0) = \pi(\theta) L(\theta; \mathbf{x}_0)$. For *frequentist* inference this gives rise to unreliable fit/prediction measures, as well as false error probabilities. A false $f(\mathbf{x}; \theta) \Rightarrow$ a false sampling distribution $F_n(t)$ for any statistic

$$T_n = g(\mathbf{X}) : \quad F(t; \theta) := \mathbb{P}(T_n \leq t; \theta) = \underbrace{\int \int \cdots \int}_{\{\mathbf{x}:\, g(\mathbf{x}) \leq t;\, \mathbf{x} \in \mathbb{R}_X^n\}} f(\mathbf{x}; \theta) d\mathbf{x}.$$

That is, statistical misspecification induces a discrepancy between the *actual* and *nominal* (assumed) *error probabilities*. The surest way to draw an invalid inference is to apply a .05 significance level test when its actual – due to misspecification – type I error probability is closer to .99.[32]

Example. To illustrate the potentially devastating effects of statistical misspecification let us consider the simple normal model (Table 8.2), and evaluate the nominal and actual error probabilities in the case when (4) is false.

Instead of (4), we assume that $Corr(X_i, X_j) = \rho$, $0 < \rho < 1$, for all $i \neq j$, $i, j = 1$, ... n, focusing on the case where: $n = 100$, $\alpha = .05$ $(c_\alpha = 2.01)$.

What are the effects of such a misspecification on the relevant error probabilities of the t-test in (7) for the *hypotheses* in (6)? For $\rho = .1$, the nominal $\alpha = .05$ becomes an *actual* type I error $(\rho = .1)$ of $\alpha^* = .317$ and the discrepancy $\alpha^* - \alpha = .267$ increases as $\rho \to 1$. Similarly, the distortions in power are sizeable, ranging from $\pi^*(.01) - \pi(.01) = .266$ to $\pi^*(.3) - \pi(.3) = -.291$. Again, the discrepancy between nominal and actual power increases as $\rho \to 1$.[33]

The Pre-Eminence of Theory (PET) Perspective

The single most important contributor to the untrustworthiness of empirical evidence in economics is the methodological framework, known as the Pre-Eminence of Theory (PET) perspective, that has dominated empirical modelling in economics since Ricardo.[34] This framework asserts that empirical modelling takes the form of constructing *simple idealized models* which capture certain key aspects of the phenomenon of interest, with a view to use such models to shed light or even explain economic phenomena, as well as gain insight concerning potential alternative policies. From the PET perspective empirical modelling is strictly theory-driven with the data playing only a subordinate role in quantifying theory models (assumed to be true).[35]

The focus in current textbook econometrics is primarily on addressing the 'quantification' of theory models, notwithstanding the gap between the variables envisaged by theory and what the available data actually measure. What is esteemed is the mathematical sophistication of the inferential techniques (estimators and tests) called for in quantifying alternative theoretical models using different types of data (time series, cross-section and panel). Since the early 1950s, prestigious econometric journals are overflowing with a bewildering plethora of *estimation* techniques, and their associated asymptotic theory based on invoking mathematically convenient assumptions that are often non-testable. Even in the cases where some of these assumptions are testable, they are rarely checked against the data.

Textbook Econometrics and Untrustworthy Evidence

As a result, the incessant accumulation of asymptotic inferential results leave the practitioner none the wiser as to 'how' and 'when' to apply them to practical modelling problems with a view to learning from data about economic phenomena of interest. All the practitioners can do is apply these techniques to a variety of data hoping that occasionally some of the computer outputs will enable them to 'tell a story' and thus publish in order to survive academia. The trouble is that most of the empirical 'evidence' upon which these stories rely is *untrustworthy*, shedding no real light on economic phenomena of interest, primarily because the link between the theory and data is precarious at best.

A crucial implication of the textbook econometric PET perspective is that by making probabilistic assumptions about error (shock) terms, the emphasis is placed on the *least restrictive assumptions* that would justify (asymptotically) the 'quantification' technique, irrespective of whether they are testable or not. The idea is that *weaker* assumptions are less vulnerable to *misspecification*; hence the current popularity of the Generalized Method of Moments (GMM) as well as nonparametric methods. This is a flawed argument because weaker, but indirect, assumptions about the stochastic process $\{\mathbf{Z}_t, \ t \in \mathbb{N}\}$ underlying the data Z_0, are *not* necessarily less vulnerable to statistical misspecification when they largely ignore the probabilistic structure of the data. Indeed, weaker non-tested/testable assumptions are likely to undermine both the reliability and precision of inference.[36]

The textbook 'quantification of theory' perspective has influenced, not only the way statistical models are *specified* (attaching error terms to theory models), but also the M-S testing and respecification facets of modelling.

The linear regression (LR) model (Table 8.3) was made the cornerstone of textbook econometrics with all other models being viewed as modifications/extensions of this model.[37] A key issue with Table 8.3 is that the *error assumptions* (i)–(iv) provide a misleading picture of the statistical premises since:

    a.   Their validity *cannot* be directly evaluated vis-a-vis data $\mathbf{Z}_0 = (\mathbf{z}_1, ..., \mathbf{z}_n)$.

b. The list of assumptions is incomplete; there are missing assumptions.

c. The parameters $(\beta_0, \beta_1, \sigma^2)$ are not given a proper statistical parameterization.

d.   What ultimately matters is not the set of assumptions pertaining to the error term, but what they imply for the probabilistic structure of the observable process $\{Z_t,\ t\in\mathbb{N}\}$, $Z_t := (y_t, X_t)$ in terms of which the distribution of the sample $D(z_1, z_2, ..., z_n; \phi)$ and the likelihood function $L(\phi; Z_0)$ are defined.

Table 8.3: Traditional linear regression (LR) model

$$y_t = \beta_0 + \beta_1 x_t + u_t,\ t\in\mathbb{N} := (1, 2, ..., n, ...)$$

(i) $(u_t \mid X_t = x_t) \sim N(.,.),$ (ii) $E(u_t \mid X_t = x_t) = 0,$ (12)

(iii) $E(u_t^2 \mid X_t = x_t) = s^2,$ (iv) $E(utus \mid Xt = xt) = 0$, for $t \neq s, t, s \in \mathbb{N}.$

e. The *substantive* and *statistical* information are inseparably *intermingled* by viewing the theory model as the *systematic component* and attaching *probabilistic error terms* to define the *non-systematic* component.

The last of these weaknesses has contributed significantly to providing a misleading perspective on both Mis-Specification (M-S) testing and respecification. To illustrate that, consider the first M-S test to probe for departures from assumption (iv), proposed by Durbin and Watson. Their proposed test was based on assuming a particular form of autocorrelation for the error term:

(iv)* $u_t = \rho u_{t-1} + \varepsilon_t,\ |\rho| < 1,$ (13)

and framing the null and alternative hypotheses in terms of the additional parameter:

$H_0 : \rho = 0,$ vs $H_1 : \rho \neq 0.$ (14)

Ignoring the fact that $E(u_t u_s \mid X_t = x_t) \neq 0$ can take an infinite number of forms, rejection of H_0 is interpreted by textbook econometrics as providing evidence for H_1. Worse, the form of respecification recommended is to adopt H_1, by making the autocorrelation-corrected (A-C) linear regression the new maintained model.[38] This form of respecification constitutes a classic example of the fallacy of rejection: evidence against H_0 (mis)interpreted as evidence for H_1; see pp. 190–207.

An Empirical Example: The CAPM

Consider the following example from financial econometrics. The *structural model* is known as the capital asset pricing model (CAPM):[39]

$$(r_{kt} - r_{ft}) = \beta_k(r_{Mt} - r_{ft}) + \varepsilon_{kt}, k = 1, 2, ..., m, t = 1, ..., n,$$ (15)

where r_{kt} – returns of asset k, r_{MT} – market returns, r_{ft} – returns of a risk free asset. To simplify the discussion, let us focus on just one of the m equations (one for each asset). To test its validity vis-a-vis the data, the CAPM (15) is usually embedded into the linear regression model (Table 8.3):

$$y_t = \beta_0 + \beta_1 x_t + u_t, \, t = 1, 2, \dots, n, \tag{16}$$

where $y_t := (r_t - r_{ft})$, $x_t := (r_{Mt} - r_{ft})$. In light of the fact that (16) nests (15) via the parametric restriction $\beta_0 = 0$, the validity of the CAPM is often tested using the hypotheses:

$$H_0 : \beta_0 = 0, \quad \text{vs} \quad H_1 : \beta_0 \neq 0. \tag{17}$$

Data. The relevant data Z_0 come in the form of the monthly log-returns of Citigroup $(r_t = \Delta \ln P_t)$ for the period August 2000 to October 2005, with the market portfolio returns (r_{Mt}) measured by the SP500 index and the risk free asset by the three-month Treasury bill (3-Tb) rate return (r_{ft}).[40]

Estimating the statistical model (16) for Citigroup yields:

$$y_t = \underset{(.0033)}{.0053} + \underset{(.089)}{1.137} x_t + \underset{(.0188)}{\hat{u}_t}, \quad R^2 = .725, \; s = .0188, \; n = 64. \tag{18}$$

On the basis of (18) the PET assessment will be as follows:[41]

   a.   the signs and magnitudes of the estimated coefficients are in accordance with the CAPM ($\beta_0 = 0$ and $\beta_1 > 0$),

   b. β_1 is statistically significant since the t-test:
$\tau(Z_0; \beta_1) = \frac{1.137}{.089} = 12.775[.000]$, where the p-value is given in square brackets,

   c.   the CAPM restriction $\beta_0 = 0$ is *not rejected* since:
$\tau(Z_0; \beta_0) = \frac{.0053}{.0033} = 1.606[.108]$,

   d.   the goodness-of-fit is reasonably high ($R^2 = .725$), providing additional support for the CAPM.

Taken together, (a)–(d) are regarded as providing good evidence *for* the CAPM.

Such confirming evidence, however, is questionable unless one can show that the invoked inductive premises are valid for the above data. As shown in the next section, the above estimated model is seriously misspecified statistically, rendering the above inferences unreliable.

Error Statistics: A Modelling Framework

In the context of the F–N–P perspective, probability plays two interrelated roles. First, $f(z; \theta)$, $z \in \mathbb{R}_Z^n$ attributes probabilities to all legitimate events relating to the sample Z. Second, it provides all relevant error probabilities associated with any statistic (estimator, test or predictor) $T_n = g(Z)$. Arguably, the most enduring contribution of the F–N–P perspective comes in the form of rendering the errors for statistical induction ascertainable by embedding the 'material experiment' into a statistical model in terms of which the relevant error probabilities are defined, thus providing a measure of the reliability of the inference procedure.[42]

Outstanding Foundational Problems and Issues

Despite the widespread influence of the F–N–P framework in numerous applied fields, several key foundational issues and problems, bedevilling this approach since 1930s, remained unresolved. These problems include not only issues pertaining to inference, extensively discussed in the statistics literature, but also problems pertaining to modelling arising in fields where substantive information plays an important role. A partial list of these key problems is given below.

A. Foundational issues pertaining to modelling
   (i) Statistical model specification: choosing a statistical model $\mathcal{M}_\theta(\mathbf{z})$ as it relates to:
      (a) statistical information contained in data Z_0, and
      (b) substantive (subject matter) information,
   (ii) statistical adequacy: establishing the validity of a statistical premises $\mathcal{M}_\theta(\mathbf{z})$ vis-a-vis data Z_0,
   (iii) statistical model respecification: how to respecify a model $\mathcal{M}_\theta(\mathbf{z})$ when found *misspecified*, and
   (iv) substantive adequacy: how to assess the adequacy of a substantive model $\mathcal{M}_\varphi(\mathbf{z})$ in shedding light on (describing, explaining, predicting) the phenomenon of interest.

B. Foundational issues pertaining to inference
   (v) the role of pre-data (type I and II) versus post-data (p-value) error probabilities,
   (vi) safeguarding frequentist inference against the following fallacies:
      a. the fallacy of acceptance: (mis)interpreting accept H_0 (no evidence against H_0) as evidence for H_0,
      b. the fallacy of rejection: (mis)interpreting reject H_0 (evidence against H_0) as evidence for a particular H_1; e.g. statistical versus substantive significance.

As argued next, these issues and problems can be adequately addressed in the context of the 'error statistical approach', which can be viewed as a refinement/ extension of the F–N–P).[43] In particular, error statistics aims to:
   [A] *Refine* the F–N–P approach by proposing a broader framework that distinguishes the statistical from the substantive premises, with a view to addressing modelling issues like (i)–(iv) and

[B] *Extend* the F–N–P approach by supplementing it with a post-data severity assessment with a view to addressing a variety of inference problems, including (v)–(vi).[44]

The most important modelling foundational problem that has frustrated every applied field that relies on observational data relates to the role of theory and data in empirical modelling and pertains to the issues (i)–(iv) raised above. The inevitable result of foisting one's favourite theory on the data is often an estimated model $\mathcal{M}_\varphi(\mathbf{z})$ which is both *statistically* and *substantively misspecified*, but one has no principled way to distinguish between the two sources of misspecification and apportion blame:

> is the substantive subject matter information false? or
> are the *statistical premises* of inference misspecified?   (19)

The key to circumventing this *Duhemian ambiguity*[45] is to find a way to disentangle the statistical $\mathcal{M}_\theta(\mathbf{z})$ from the substantive premises $\mathcal{M}_\varphi(\mathbf{z})$ without compromising the integrity of either source of information.

What is often insufficiently realized is that behind every inference based on observed data $\mathbf{Z}_0$ in conjunction with a substantive (scientific) model $\mathcal{M}_\varphi(\mathbf{z})$, there is a distinct statistical model $\mathcal{M}_\theta(\mathbf{z})$ (often implicit) providing the inductive premises of inference. Separating the two enables one to establish statistical adequacy first in order to secure the reliability of inference procedures, and then proceed to assess the substantive hypotheses of interest reliably. In addition, sound probabilistic foundations for the notion of a statistical model deal effectively with the crucial problem of distinguishing between real regularities and artefacts in data.

Bridging the Gap between Theory and Data

Error statistics[46] can be viewed as a refinement/extension of the F–N–P approach to frequentist statistics, which proposes a methodological framework that can be used to address the foundational problems in (i)–(vi), as well as the methodological problems (i)–(vi) raised by Mill[47] and Keynes.[48] The key pillars of error statistics are the following:

(A) From theory T to testable hypotheses h: fashioning an abstract and idealized theory T into a *structural (substantive) model* $\mathcal{M}_\varphi(\mathbf{z})$ that is *estimable* in light of data $\mathbf{Z}_0$, with h representing the substantive hypotheses of interest framed in the context of $\mathcal{M}_\varphi(\mathbf{z})$.

To address this problem, error statistics proposes a sequence of interconnected models[49] aiming to bridge the gap between theory and data. It is interesting to note that the idea of relating theory to data using a sequence of models was proposed by Suppes[50] whose hierarchy of models *theory*,[51] *experimental* and *data models* shares certain affinities and correspondence with the sequence models *theory, structural (estimable)* and *statistical*, by Spanos.[52] However, there are also crucial differences because the Suppes hierarchy focuses primarily on experimental data where controls and experimental design techniques can be used to provide a direct link to the theory and data models; no such tools are available when modelling non-experimental data. Some of these differences are brought out below.

From the theory side, one constructs a theory model (a mathematical formulation), say $\mathcal{M}_\psi(\mathbf{z}; \xi)$ which often includes *latent* variables ξ. In connecting $\mathcal{M}_\psi(\mathbf{z}; \xi)$ to the available data $\mathbf{Z}_0$ one needs to transform the theory model into an estimable (in light of data $\mathbf{Z}_0$) form, the substantive (structural) model $\mathcal{M}_\varphi(\mathbf{z})$. The construction of a structural model is particularly taxing in economics because one needs to reflect on the huge gap between the circumstances envisaged by the theory and its concepts (intentions, plans), and the actual phenomenon of interest giving rise to the available data.[53]

From the data side, the statistical information (chance regularity patterns exhibited by data) is distilled by a *statistical model* $\mathcal{M}_\theta(\mathbf{z})$ with a view to meeting two interrelated aims:

(a) to account for the chance regularities in data Z_0 by choosing a probabilistic structure for the generic stochastic process $\{\mathbf{Z}_t, \ t \in \mathbb{N}\}$ so as to render Z_0 a 'truly typical realization' thereof, and

(b) to parameterize ($\theta \in \Theta$) this probabilistic structure in an attempt to specify $\mathcal{M}_\theta(\mathbf{z})$ in such a way so as to embed (parametrically) $\mathcal{M}_\varphi(\mathbf{z})$ in its context via restrictions of the form $G(\theta, \varphi) = 0, \theta \in \Theta, \varphi \in \Phi$, relating the statistical and substantive parameters.

In light of these identification restrictions (see Figure 8.3) one might be interested in specific substantive hypotheses of interest h that can now be framed in terms of the statistical parameters θ, say:

$$H_0 : h(\theta) = 0, \quad \text{vs} \quad H_1 : h(\theta) \neq 0, \theta \in \Theta. \tag{20}$$

(B) **From raw data to reliable evidence**: transforming a finite and incomplete set of raw data Z_0 – containing uncertainties, impurities and noise – into reliable 'evidence' summarized by a statistically adequate model $\mathcal{M}_\theta(\mathbf{z})$ pertinent for appraising the substantive hypotheses of interest h.

This takes the form of establishing the statistical adequacy of $\mathcal{M}_\theta(\mathbf{z})$ using thorough *(M-S) testing* (validation), guided by discerning graphical techniques and effective probing strategies, and respecification in cases where the original model is found wanting. A statistically adequate model $\mathcal{M}_\theta(\mathbf{z})$ secures the reliability of statistical inferences based on it by ensuring that the *actual error probabilities* approximate closely enough the *assumed* (nominal) ones. It is important to note that goodness-of-fit/prediction is neither necessary nor sufficient for statistical adequacy.[54]

(C) **Using reliable evidence to appraise substantive hypotheses**: assessing whether Z_0 provides evidence for or against $h(\theta) = 0$.

A statistically adequate model $\mathcal{M}_\theta(\mathbf{z})$ summarizes the statistical information in the data to provide the reliable evidence against which the theory in question can be confronted. Error statistics calls for probing the substantive hypotheses of interest in (20) only in the context of a statistically adequate model $\mathcal{M}_\theta(\mathbf{z})$ (to ensure the reliability of the test in question). Moreover, it proposes extending the N–P accept/reject result to providing evidence for or against $h(\theta) = 0$ using the *severity reasoning*. The latter establishes the warranted discrepancy from the null and enables one to address the crucial fallacies of acceptance and rejection.[55]

Similarly, one can initiate the process of assessing the *substantive adequacy* of $\mathcal{M}_\varphi(\mathbf{z})$ by testing the over-identifying restrictions in the form of the hypotheses:

$$H_0 : G(\theta, \varphi) = 0, \quad \text{vs} \quad H_1 : G(\theta, \varphi) \neq 0, \text{ for } \theta \in \Theta, \varphi \in \Phi.$$

When empirically valid, the learning from data is passed from θ to j, rendering $\mathcal{M}_\varphi(\mathbf{z})$ both statistically and substantively meaningful.[56] $\mathcal{M}_{\hat{\varphi}}(\mathbf{z})$ can now provide the basis for shedding light on (describing, explaining, predicting) the phenomenon of interest.

Figure 8.3 is a simplified version of the original Diagram 1.2, which was published elsewhere,[57] and aims to bring out the statistical induction dimension of the methodological framework.

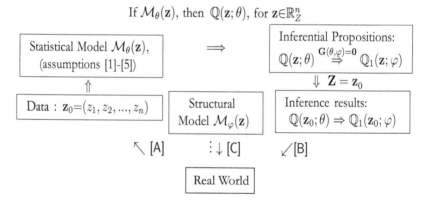

Figure 8.3: Model-based frequentist statistical induction.

Probing for and Eliminating/Controlling Errors

One of the key advantages of bridging the gap between theory and data in terms of a sequence of interrelated models (Figure 8.3) is that it makes the framing of potential errors during the modelling process more transparent.

I. **Errors in theory model construction:** the mathematical framing of an abstract theory T into a theory model $\mathcal{M}_\psi(\mathbf{z}; \xi)$ might turn out to be inappropriate or incomplete.

II. **Errors in eliminating latent variables:** transforming $\mathcal{M}_\psi(\mathbf{z}; \xi)$ into a structural model $\mathcal{M}_\varphi(\mathbf{z})$ to render it estimable with data $\mathbf{Z}_0$.

III. **Inaccurate data errors:** data $\mathbf{Z}_0$ are marred by systematic errors imbued by the collection/compilation process. Such systematic errors are likely to distort the statistical regularities and give rise to misleading inferences.[58]

IV. **Incongruous measurement errors:** data $\mathbf{Z}_0$ do not adequately quantify the concepts envisioned by the theory. This, more than the other substantive sources of error, is likely to be the most serious factor in ruining the trustworthiness of empirical evidence.[59]

V. **Statistical misspecification errors:** one or more of the probabilistic assumptions of the statistical model $\mathcal{M}_\theta(\mathbf{z})$ is invalid for data $\mathbf{Z}_0$.

VI. **Substantive inadequacy errors:** the circumstances envisaged by the theory in question differ 'systematically' from the actual data-generating mechanism underlying the phenomenon of interest. This inadequacy can easily arise from impractical *ceteris paribus* clauses, missing confounding factors, false causal claims, etc.[60]

Statistical versus Substantive Premises of Inference

The confusion between statistical and substantive assumptions permeates the whole econometric literature. A glance at the complete specification of the linear regression model in econometric textbooks reveals that along the probabilistic assumptions pertaining to the observable processes (made via the error term) underlying the data in question, there are substantive assumptions concerning omitted variables,[61] measurement errors and non-simultaneity.[62]

The key question is: how can one disentangle the two types of assumptions?

Error statistics views empirical modelling as a *piecemeal process* that relies on distinguishing between the *statistical* $\mathcal{M}_\theta(\mathbf{z})$ and *substantive models* $\mathcal{M}_\varphi(\mathbf{z})$, clearly delineating the following two questions:

a. statistical adequacy: does $\mathcal{M}_\theta(\mathbf{z})$ account for the chance regularities in $\mathbf{Z}_0$?

b. substantive adequacy: is model $\mathcal{M}_\varphi(\mathbf{z})$ adequate as an explanation (*causal* or otherwise) of the phenomenon of interest?

As sketched above, the substantive premises stem from the theory or theories under consideration. The more difficult question is: where do the statistical premises come from?

The fashioning of the statistical model (premises) $\mathcal{M}_\theta(\mathbf{z})$ begins with a given data $\mathbf{Z}_0$, irrespective of the theory or theories that led to the choice of $\mathbf{Z}_0$. Once selected, data $\mathbf{Z}_0$ take on 'a life of its own' as a particular *realization* of an underlying generic stochastic process $\{\mathbf{Z}_t, \ t \in \mathbb{N}\}$. The link between data $\mathbf{Z}_0$ and the process $\{\mathbf{Z}_t, \ t \in \mathbb{N}\}$ is provided by the key question: what probabilistic structure pertaining to the process $\{\mathbf{Z}_t, \ t \in \mathbb{N}\}$ would render data $\mathbf{Z}_0$ a *truly typical realization* thereof?

Adopting a *statistical perspective*, one views data $\mathbf{Z}_0$ as a realization of a generic (vector) stochastic process $\{\mathbf{Z}_t, \ t \in \mathbb{N}\}$, regardless of what the variables $\mathbf{Z}_t$ measure substantively, thus separating the 'statistical' from the 'substantive' information.

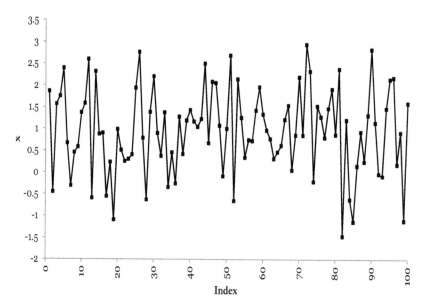

Figure 8.4: NIID realization.

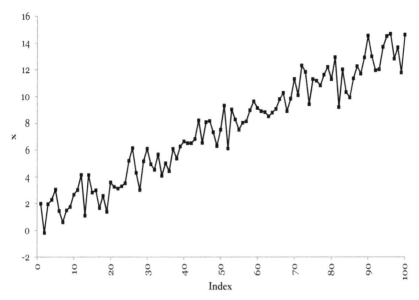

Figure 8.5: NI, mean-heterogeneous.

Example. What does a 'typical realization' of Normal, Independent and Identically Distributed (NIID) process $\{Z_t, \ t\in\mathbb{N}\}$ look like? It looks like Figure 8.4, but not like Figure 8.5.[63]

A pertinent answer to the 'typicality' question provides the relevant probabilistic structure for $\{Z_t, \ t\in\mathbb{N}\}$. The next step is to choose a parameterization $\theta \in \Theta$, based on this probabilistic structure, to define the relevant *statistical model* $\mathcal{M}_\theta(\mathbf{z})$.

Example 1. For the data in Figure 8.4, an appropriate model is the simple normal:

$$\mathcal{M}_\theta(\mathbf{z}): Z_t \backsim \mathsf{NIID}(\mu, \sigma^2), \ \theta:=(\mu, \sigma^2)\in\mathbb{R}\times\mathbb{R}_+, \ z_t\in\mathbb{R}, \ t\in\mathbb{N}$$

Example 2. For Figure 8.5, an appropriate model is the normal, mean-heterogeneous:

$$\mathcal{M}_\theta(\mathbf{z}): Z_t \backsim \mathsf{NI}(\mu_0 + \mu_1 t, \sigma^2), \ \theta:=(\mu_0, \mu_1, \sigma^2)\in\mathbb{R}^2\times\mathbb{R}_+, \ z_t\in\mathbb{R}, \ t\in\mathbb{N}$$

The particular parameterization $\theta \in \Theta$ giving rise to $\mathcal{M}_\theta(\mathbf{z})$ is chosen with a view to embed (nest) the *substantive model* $\mathcal{M}_\varphi(\mathbf{z})$ in its context, e.g. via certain restrictions

$\mathbf{G}(\theta, \varphi) = 0, \ \varphi \in \Phi.$[64]

A statistical model of particular interest in economics is the normal/linear regression (LR) model, which from this perspective can be viewed as a parameterization of an observable NIID process $\{\mathbf{Z}_t := (y_t, X_t), \ t \in \mathbb{N}\}$, stemming from the reduction:

$$D(\mathbf{Z}_1, \mathbf{Z}_2, ..., \mathbf{Z}_n; \phi) \overset{\text{I}}{=} \prod_{t=1}^{n} D_t(\mathbf{Z}_t; \varphi(t)) \overset{\text{IID}}{=} \prod_{t=1}^{n} D(\mathbf{Z}_t; \varphi) = \prod_{t=1}^{n} D(y_t | \mathbf{x}_t; \varphi_1) D(\mathbf{X}_t; \varphi_2)$$

where normality of $D(Z_t; \varphi)$ ensures that the normal LR can be specified exclusively in terms of the conditional distribution $D(y_t | x_t; \varphi_1)$, as given in Table 8.4. Note that the multivariate normal model, the principal components and factor analysis models can be viewed as alternative parameterizations of the same NIID process $\{\mathbf{Z}_t := (y_t, X_t), \ t \in \mathbb{N}\}$.

Table 8.4: Normal/linear regression model

Statistical GM: $y_t = \beta_0 + \beta_1 x_t + u_t, t \in \mathbb{N}$,
(1) Normality: $\quad (y_t | X_t = x_t) \sim N(.,.)$,
(2) Linearity: $\quad E(y_t | X_t = x_t) = \beta_0 + \beta_1 x_t$,
(3) Homosk/city: $\quad Var(y_t | X_t = x_t) = \sigma^2 > 0$,
(4) Independence: $\quad \{E(y_t | X_t = x_t), t \in \mathbb{N}\}$ independent,
(5) t-invariance: $\quad \theta := (\beta_0, \beta_1, \sigma^2)$ do not change with t.
$\beta_0 = E(y_t) - \beta_1 E(x_t), \ \beta_1 = \frac{Cov(x_t, y_t)}{Var(x_t)}, \sigma^2 = Var(y_t) - \beta_1 Cov(y_t, x_t)$

Comparing the two specifications (Table 8.3 versus Table 8.4), it is clear that there is an equivalence between assumptions (i)–(iii) and (1)–(3), (iv) is strengthened to independence, and assumption (5) is missing from Table 8.3. In contrast to Table 8.3, Table 8.4 provides a complete, internally consistent and testable set of probabilistic assumptions, without which model validation will be hopeless.

M-S Testing: The Error-Statistical Perspective

Adequate understanding of M-S testing requires distinguishing it from N–P testing. The key difference is that N–P testing constitutes probing *within* the boundaries of $\mathcal{M}_\theta(\mathbf{z}) = \{f(\mathbf{z}; \theta), \theta \in \Theta\}$, by framing the hypotheses in terms of a partition of the parameter space Θ:

$$H_0: \theta \in \Theta_0 \quad \text{vs} \quad H1: \theta \in \Theta_1, \text{ where } \Theta_0 \cup \Theta_1 = \Theta \text{ and } \Theta_0 \cap \Theta_1 = \emptyset.$$

There are only two types of errors because it is assumed that the statistical adequacy of $\mathcal{M}_\theta(\mathbf{z})$ has been secured. In contrast, M-S testing constitutes probing *outside* the boundaries of $\mathcal{M}_\theta(\mathbf{z})$, and thus no such adequacy can be secured.[65] The generic hypotheses for M-S testing take the form:

$$H: f^*(\mathbf{z}) \in \mathcal{M}_\theta(\mathbf{z}) \quad \text{vs} \quad \overline{H}: f^*(\mathbf{z}) \in [\mathcal{P}(\mathbf{z}) - \mathcal{M}_\theta(\mathbf{z})],$$

where $f^*(z)$ is the '*true*' distribution of the sample, and $\mathcal{P}(z)$ denotes the set of all possible statistical models that could have given rise to z_0. This difference raises a number of conceptual and technical issues, including the following.

1. How to particularize $[\mathcal{P}(z) - \mathcal{M}_\theta(z)]$ to construct M-S tests. A most inefficient way to do this is to attempt to probe $[\mathcal{P}(z) - \mathcal{M}_\theta(z)]$ one model at a time, $\mathcal{M}_i(z), i = 1, 2 ...,$ which is a hopeless task. A much more efficient way that enables the elimination of an *infinite* number of alternatives at a time, is to construct M-S tests by modifying the original tripartite (distribution, dependence, heterogeneity) partitioning of $\mathcal{P}(z)$ that gave rise to $\mathcal{M}_\theta(z)$ in particular directions of departures gleaned from exploratory data analysis to construct alternative models $\mathcal{M}_1(z)$ or broad subsets of $[\mathcal{P}(z) - \mathcal{M}_\theta(z)]$.[66]

2. In contrast to N–P testing, the more serious error in M-S testing is the type II error (accepting H_0 when false). This is an instance of the *fallacy of acceptance*: (mis)interpreting that accepting H_0 (i.e. no evidence against H_0) as evidence for it. Falsely rejecting $\mathcal{M}_\theta(z)$ as misspecified can be remedied at the respecification stage; no such recourse exists for falsely accepting $\mathcal{M}_\theta(z)$ as statistically adequate. Hence, a good strategy is to apply multiple M-S tests to each assumption (or combination), ensuring that some of these tests have high enough power.

3. Securing the effectiveness/reliability of the diagnosis. In M-S testing the objective is to probe as *broadly* away from the null as possible, and thus omnibus tests with low power but broad (local) probing capacity have an important role to play when combined with high power directional tests. The reliability of the diagnosis is enhanced by:

i.   *Astute ordering* of M-S tests so as to exploit the interrelationship among the model assumptions with a view to 'correct' each other's diagnosis, and

ii.   *Joint M-S tests* (testing several assumptions simultaneously) designed to minimize the maintained model assumptions. This probing strategy aims to distinguish between the different sources of misspecification. For the LR model (Table 8.4), one can construct joint M-S tests by framing potential departures in terms of how they might alter the first two conditional moments of $\mathcal{M}_\theta(z)$.

This yields *auxiliary regressions* based on the studentized residuals:

$$\widehat{v}_t = \sqrt{n}(y_k - \widehat{\beta}_0 - \widehat{\beta}_1 x_t)/s:$$

$$\widehat{v}_t = \gamma_{10} + \gamma_{11}x_t + \underset{[5]}{\underbrace{\gamma_{12}t}} + \underset{[2]}{\underbrace{\gamma_{13}x_t^2}} + \underset{[4]}{\underbrace{\gamma_{14}x_{t-1} + \gamma_{15}y_{t-1}}} + \varepsilon_{1t},$$

$$\widehat{v}_t^2 = \gamma_{20} + \underset{[3]}{\underbrace{\gamma_{21}x_t}} + \underset{[5]}{\underbrace{\gamma_{22}t}} + \underset{[3]}{\underbrace{\gamma_{23}x_t^2}} + \underset{[4]}{\underbrace{\gamma_{24}x_{t-1}^2 + \gamma_{25}y_{t-1}^2}} + \varepsilon_{2t},$$

If the residuals are *non-systematic* – the assumptions of $\mathcal{M}_\theta(z)$ are valid for data z_0 – the added terms will be statistically insignificant. Otherwise departures from model assumptions are indicated.

4. Increased vulnerability to the *fallacy of rejection*: (mis)interpreting reject H_0 (evidence against H_0) as evidence *for* H_1. M-S testing is more akin to Fisher's significance testing where the alternative, although implicitly present, should never be accepted without further testing.

Let us return to the Durbin–Watson (D–W) test for assessing assumption (4) (Table 8.1) whose hypotheses in (14) can be viewed as stemming from the fact that for:

$$\mathcal{M}_0(\mathbf{z}): \quad y_t = \beta_0 + \beta_1 x_t + u_t, \qquad\qquad u_t \backsim \text{NIID}(0, \sigma_u^2)$$
$$\mathcal{M}_1(\mathbf{z}): \quad y_t = \beta_0 + \beta_1 x_t + u_t, \ u_t = \rho u_{t-1} + \varepsilon_t, \quad \varepsilon_t \backsim \text{NIID}(0, \sigma_\varepsilon^2)$$

$\mathcal{M}_1(\mathbf{z})|_{\rho=0} = \mathcal{M}_0(\mathbf{z})$. In the traditional textbook approach, when the null is rejected the recommendation is to adopt $\mathcal{M}_1(\mathbf{z})$, ignoring the fact that the D–W test can provide good evidence *against* $\mathcal{M}_0(\mathbf{z})$, but no evidence *for* $\mathcal{M}_1(\mathbf{z})$; the rejection could have been due to any one of an infinite number of alternatives in $[\mathcal{P}(\mathbf{z}) - \mathcal{M}_\theta(\mathbf{z})]$. Securing evidence for $\mathcal{M}_1(\mathbf{z})$ requires one to test its own assumptions.

Despite the blatancy of the fallacy of rejection, the econometric literature ennobled it into the *pre-test bias* problem,[67] adding another misguided problem to textbook econometric modelling.[68]

Validating the Statistical Premises of the CAPM

Thorough M-S testing reveals (see Table 8.5) that the estimated model (18) is statistically misspecified; assumptions (1), (3)–(5) are invalid.

Table 8.5: M-S testing results

Normality: $D'AP = 1.169[.557]^\dagger$	Linearity: $F(1, 61) = .468[.496]$
Homoskedasticity: $F(2, 59) = 4.950[.010]^*$	Independence: $F(1, 59) = 6.15[.016]^*$
t-invariance: $F_\beta(2, 60) = 4.611[.014]^*$	

Indeed, one can easily see why the above statistical model is so seriously misspecified by just looking at the t-plots of the data, which are clearly not exhibiting realizations of NIID processes (Figure 8.4); both t-plots exhibit mean and variance heterogeneity as well as temporal dependence in the form of cycles.

This implies that the estimated model cannot be used as a basis of reliable inferences to assess the validity of the substantive questions of interest. To illustrate the serious errors such a strategy can give rise to, let us consider the question of confounding in the context of (18) by posing the question of whether last period's excess returns of General Motors is a relevant variable in explaining y_{1t-1}- excess returns of Citigroup. The augmented model yields the following:

$$y_t = \underset{(.0031)}{.0055} + \underset{(.114)}{1.131} x_t - \underset{(.056)}{.144} y_{1t-1} + \underset{(.0182)}{\widehat{v}_t}, \quad R^2 = .747, \ s = .0182,$$

The t-test result $(\tau(\mathbf{Z_0})=\frac{.144}{.056}=2.571)$, when taken *at face value*, indicates that the answer is yes! However, this is highly misleading because any data series that exhibits certain regularity patterns is likely to appear significant when added to the original misspecified model, including generic trends and lags:

$$y_t = \underset{(.0116)}{.0296} + \underset{(.119)}{1.134x_t} - \underset{(.065)}{.134t} + \underset{(.083)}{.168t^2} + \underset{(.0184)}{\widehat{v}_t} \;,\quad R^2{=}.745, \; s{=}.0184,$$

$$y_t = \underset{(.003)}{.0034} + \underset{(.094)}{1.24x_t} - \underset{(.071)}{.175y_{t-1}} + \underset{(.0181)}{\widehat{v}_t} \;,\quad R^2{=}.75, \; s{=}.0181.$$

Indeed, this can be used to explain the 'apparent success' of the Fama and French multifactor model:[69]

$$\mathcal{M}_\psi(\mathbf{z}): \quad (r_{it}-r_{ft}) = \beta_{0i}+\beta_{1i}(r_{Mt}-r_{ft})+\textstyle\sum_{i=2}^{p}\beta_{ii}f_{it}+\varepsilon_{it}, \;\; i{=}1,2,...,m, \;\; t{=}1,...,n,$$

where f_i, $i = 1, 2, ..., p$ are omitted relevant factors, being nothing more than misguided substantive respecifications stemming from a misspecified statistical model.

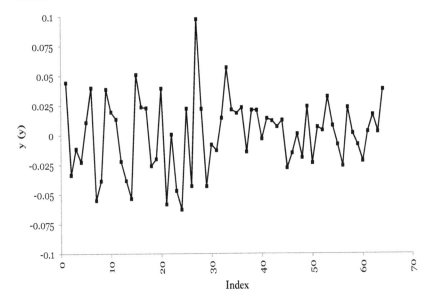

Figure 8.6: t-plot of CITI excess returns.

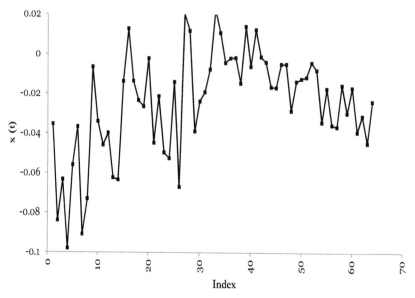

Figure 8.7: t-plot of market excess returns.

The question that naturally arises at this stage, is what is one supposed to do next? Three things are clear.

First, no evidence for or against a substantive claim (theory) can be secured on the basis of a statistically misspecified model. *Second*, the so-called refinements/extensions of the CAPM[70] are misguided attempts driven by unreliable inference results! This is also the case in various other applied subfields in economics. *Third*, before any reliable testing of the CAPM can be applied one needs to secure the statistical adequacy; the validity of model assumptions vis-a-vis data Z_0.

Statistical versus Substantive Adequacy

The distinction between statistical and substantive adequacy can be used to shed light on several confusions that permeate the current textbook approach to econometric modelling, including the following:

1. Statistical adequacy is the price for using reliable statistical procedures to learn from data; it ensures the reliability of inference by securing the closeness of the *actual and nominal error probabilities*. Learning from data depends crucially on establishing a *sound link* between the process generating the data Z_0 and the assumed $\mathcal{M}_\theta(z)$, by securing statistical adequacy. Statistical misspecification also calls into question other criteria used by the Pre-Eminence of Theory (PET) perspective to evaluate the appropriateness of an estimated model:

   (i) statistical: goodness-of-fit/prediction, statistical significance
   (ii) substantive: theoretical meaningfulness, explanatory capacity
   (iii) pragmatic: simplicity, generality, elegance

because, directly or indirectly, they invoke inferences. Any inference based on a statistically misspecified model is likely to be unreliable, yielding untrustworthy evidence.

2. Statistical adequacy depends only on $\mathcal{M}_\theta(\mathbf{z})$ and $\mathbf{Z}_0$ and can be established independently by different practitioners using thorough M-S testing; it is well-defined and objective. Hence, to paraphrase Keynes: 'If 70 well-trained econometricians were shut up in 70 separate rooms with the same $\mathcal{M}_\theta(\mathbf{z})$ and $\mathbf{Z}_0$, when they emerge they should have the same misspecification diagnosis.'[71] Such a claim cannot be made in relation to establishing substantive adequacy because the search is more open-ended, in the sense that there is an infinity of scenarios one could potentially probe for.

3. Statistical adequacy is necessary but not sufficient for substantive adequacy. For substantive adequacy one needs to probe for additional *potential errors* vis-a-vis the phenomenon of interest, impractical *ceteris paribus* clauses, intentions versus realizations, omitted effects, false causal claims, etc.

Example. The textbook discussion of the 'omitted variables' bias and inconsistency problem is mired in confusion. A closer look at the argument reveals that it has nothing to do with statistical misspecification. Despite the allusions to biases and inconsistencies of estimators, it is an issue pertaining to the *substantive adequacy* of a structural model $\mathcal{M}_\varphi(\mathbf{z})$.[72]

From this perspective, Keynes's thesis that one needs to account for all the relevant contributing factors (including non-measurable ones) at the outset, amounts to claiming that to *learn anything* from data one must *know everything* at the outset. Despite the fact that the substantive (structural) model $\mathcal{M}_\varphi(\mathbf{z})$ may always come up short in fully capturing or explaining a highly complex phenomenon of interest, a statistical model $\mathcal{M}_\theta(\mathbf{z})$ may be entirely adequate to reliably test and evaluate substantive questions of interest, including the empirical adequacy of $\mathcal{M}_\varphi(\mathbf{z})$.

4. Good fit is neither necessary nor sufficient for statistical adequacy. It is often claimed that a 'good' model is one that 'fits the data or predicts well'. This is highly misleading because *good fit/prediction* is neither necessary nor sufficient for the statistical adequacy of $\mathcal{M}_\theta(\mathbf{z})$.[73] As mentioned above, statistical misspecification also undermines the reliability of all statistics used to measure goodness-of-fit/prediction, including the R^2 and the Mean Square Error (MSE). This problem is illustrated by Mayo and Spanos[74] using a simple linear regression (LR) $y_t = \beta_0 + \beta_1 x_t + u_t$, where the data $\mathbf{z}_0 := \{(y_t, x_t), t = 1, 2 \ldots n\}$ exhibit both heterogeneity (trends in mean) and temporal dependence (cycles) rendering the LR statistically misspecified because assumptions (2)–(5) (Table 8.4) are likely to be invalid.[75] If the OLS estimates and related statistics are taken at face value, the t-ratios (270.0, 79.1) for the coefficients (β_0, β_1) indicate strong statistical significance and the $R^2 = .995$ indicates almost perfect fit. However, none of

these statistics is trustworthy. When a statistically adequate model, in the form of a dynamic linear regression model with a trend is estimated, the two variables are shown to be totally unrelated. Indeed, statistical misspecification can be used to account for most 'spurious results'.[76]

On the other hand, good fit might be relevant for *substantive adequacy* in the sense that for an empirically valid structural model $\mathcal{M}_\varphi(\mathbf{z})$ goodness-of-fit can provide a measure of its comprehensiveness vis-a-vis the phenomenon of interest.

5. The slogan: 'all models are false, but some are useful', attributed to George Box, is based on confusing substantive and statistical inadequacy. There is nothing wrong with constructing a simple, abstract and idealized theory-model $\mathcal{M}_\varphi(\mathbf{z})$ aiming to capture key features of the phenomenon of interest. The problem arises when the data Z_0 are given a *subordinate* role, that of 'quantifying' $\mathcal{M}_\varphi(\mathbf{z})$ that (i) ignores the probabilistic structure of the implicit statistical model $\mathcal{M}_\theta(\mathbf{z})$ and gives rise to untrustworthy evidence, and (ii) does not carry out reliable testing of whether $\mathcal{M}_\varphi(\mathbf{z})$ does, indeed, capture the key features of the phenomenon of interest. A substantive model $\mathcal{M}_\varphi(\mathbf{z})$ may always come up short in fully capturing or explaining a phenomenon of interest (e.g. not realistic enough), but a statistical model $\mathcal{M}_\theta(\mathbf{z})$ may be entirely satisfactory (when it is statistically adequate) to reliably test substantive questions of interest, including assessing the appropriateness of $\mathcal{M}_\varphi(\mathbf{z})$. That is, even unrealistic models can give rise to learning from data, but they need to be statistically adequate.

6. Empirical modelling in the social sciences, where the actual data generating mechanisms are highly complex and the available data are (often) observational, calls for *more* not *less* reliance on statistical methods. For instance, a statistically adequate $\mathcal{M}_\theta(\mathbf{z})$ could play a more crucial role in *guiding the search* for substantively adequate explanations (theories) by delineating 'what there is to explain'. Kepler's 'law' for the elliptical motion of the planets was the empirical regularity that guided Newton toward a substantively adequate model (universal gravitation) sixty years after it was first established.[77]

Post-Data Severity Evaluation

Error statistics provides an important extension of the F–N–P approach to frequentist inference in the form of a *post-data evaluation* of the N–P accepting (rejecting) results using severe testing reasoning. This provides a link between frequentist results like accepting/rejecting the null hypothesis or a large/small p-value and an *evidential interpretation* (evidence for or against a hypothesis) in the form of the smallest/largest discrepancy g from the null warranted with data x_0.[78]

To illustrate the severity evaluation, consider testing the following hypotheses:

$$H_0 : \theta = \theta_0 \text{ vs } H_1 : \theta > \theta_0, \text{ where } \theta_0 = \tfrac{18}{35} \simeq .5143,$$

in the context of the simple Bernoulli model (Table 8.1) for $\alpha = .01 \Rightarrow c_\alpha = 2.326$. It can be shown that the test defined by $T_\alpha^> := \{d(\mathbf{X}), C_1^>(\alpha)\}$:

$$d(\mathbf{X}) = \frac{\sqrt{n}(\overline{X}_n - \theta_0)}{\sqrt{\theta_0(1-\theta_0)}} \overset{H_0}{\backsim} \text{Bin}(0, 1; n), \quad C_1^>(\alpha) = \{\mathbf{x}: d(\mathbf{x}) > c_\alpha\}, \quad (21)$$

is an α-level, Uniformly Most Powerful (α-UMP) test.[79] Applying this test to the above data $\mathbf{x}_0$, where $n = 9{,}869$, boys $= 5{,}152$, girls $= 4{,}717$ yields:

$$d(\mathbf{x}_0) = \frac{\sqrt{9869}(\frac{5152}{9869} - \frac{18}{35})}{\sqrt{\frac{18}{35}(1 - \frac{18}{35})}} = 1.541, \quad p_>(\mathbf{x}_0) = \mathbb{P}(d(\mathbf{X}) > 1.541; H_0) = .062.$$

Hence, the null hypothesis is *accepted*, and the p-value indicates that H_0 wouldn't have been rejected even if $\alpha = .05$ were to be chosen. Does the acceptance mean that there is evidence for $\theta_0 = \frac{18}{35}$? Not necessarily! As argued above, acceptance (rejection) of H_0 is vulnerable to the fallacy of acceptance (rejection).

An effective way to circumvent such fallacious inferences is to use the post-data severity evaluation.[80] A hypothesis H passes a severe test T_α with data $\mathbf{x}_0$ if:

(S-1) $\mathbf{x}_0$ accords with H, and

(S-2) with very high probability, test T_α would have produced a result that accords less well with H than $\mathbf{x}_0$ does, if H were false.

In light of the fact that for the above example $d(\mathbf{x}_0) = 1.541 > 0$, the relevant inferential claim is of the form $\theta > \theta_1 = \theta_0 + \gamma$, and the 'discordance' condition (S-2) of severity gives rise to:

$$SEV(T_\alpha^>; \theta > \theta_1) = \mathbb{P}(\mathbf{x}: d(\mathbf{x}) \leq d(\mathbf{x}_0); \theta = \theta_1), \quad (22)$$

whose evaluation is based on the sampling distribution:

$$d(\mathbf{X}) = \frac{\sqrt{n}(\hat{\theta}_n - \theta_0)}{\sqrt{\theta_0(1-\theta_0)}} \overset{\theta = \theta_1}{\backsim} \text{Bin}(\delta(\theta_1), V(\theta_1); n), \quad \text{for } \theta_1 > \theta_0, \quad (23)$$

where $\delta(\theta_1) = \frac{\sqrt{n}(\theta_1 - \theta_0)}{\sqrt{\theta_0(1-\theta_0)}} \geq 0$, $V(\theta_1) = \frac{\theta_1(1-\theta_1)}{\theta_0(1-\theta_0)}$, $0 < V(\theta_1) \leq 1$.

Table 8.6 reports the evaluation of the severity for several key discrepancies, including $\theta_0 \cong .5143$ (first conjectured by Nicholas Bernoulli) and the substantively determined value: $\theta^\bullet = .5122$; In human biology[81] it is commonly accepted that the sex ratio at birth is approximately 105 boys to 100 girls, yielding $\theta^* = \frac{105}{205} = .5122$. The severity evaluates in Table 8.4 indicate that for a high enough threshold probability, say .90, the smallest warranted discrepancy associated with the claim:

$\theta > \theta_1 = \theta_0 + \gamma$, is $\gamma \leq .5156$.

Table 8.6: Accept H_0: $\theta = \frac{18}{35}$ vs H_1: $\theta > \frac{18}{35}$.

γ	Relevant claim $\theta > \theta_1 = [.5143+\gamma]$,	Severity $SEV(T_\alpha^>; \mathbf{x}_0; \theta > \theta_1)$
−.0043	$\theta > 0.510$	.992
−.0033	$\theta > 0.511$	.986
−.0021	$\theta > 0.5122$	.975
−.0013	$\theta > 0.513$	.964
−.00053	$\theta > 0.51376$	.950
−.00003	$\theta > 0.514$	.945
.00000	$\theta > 0.5143$	.938
.0007	$\theta > 0.515$	.919
.0013	$\theta > 0.5156$	.900
.0017	$\theta > 0.516$	.885
.0027	$\theta > 0.517$	.842
.0037	$\theta > 0.518$	.789
.0057	$\theta > 0.520$	.657

Severity and learning from data. The inferential claims associated with the post-data severity evaluation enable one to get a lot more accurate information pertaining to what the data say about θ. Taking an overly simplistic perspective, one can argue that these evaluations provide very good evidence (of severity .9 or higher) for a tiny range of values:

$$\text{from } [0 \leq \theta \leq 1] \xRightarrow{SEV(T_\alpha^>;\mathbf{x}_0;\theta>\theta_1)} [.51 \leq \theta \leq .5156], \tag{24}$$

with the severity evaluation gradating them. It is important to notice that the point estimate $\hat{\theta}_n(\mathbf{x}_0) = .522$ lies outside (24), and the .95 observed CI [.51215, .53186] hardly overlaps with this range of values! Moreover, the above interval should not be interpreted as attaching probabilities to different values of θ, but as post-data $(\mathbf{x}_0)$ error probabilities, based on the test $T_\alpha^>$ attached to particular inferential claims $(\theta > \theta_1)$.

Summary and Conclusions

A widely held view in economics is that the current untrustworthiness of empirical evidence and the inability to forecast economic phenomena is largely due
to the fact that the economy is too complicated and the resulting data too heterogeneous to be amenable to statistical modelling. This chapter has argued
that whether economic data are amenable to statistical modelling is largely an
empirical issue and the current untrustworthiness is mainly due to an inadequate
implementation of statistical modelling and inference. The latter stems from two
main sources: (i) commingling the statistical and substantive premises of inference that leads to numerous confusions, and (ii) neglecting the validation of the
latter, which gives rise to unreliable inferences.

A case was made that a more appropriate implementation of frequentist
statistics can be achieved using the error statistical perspective, a refinement/
extension of the F–N–P approach, that attains *learning from data* about phenomena of interest by addressing several foundational problems, including
(i)–(ii) above, in a way that gives rise to reliable and effective procedures to
establish trustworthy evidence.

The key to disentangling the statistical from the substantive premises is to
adopt a purely probabilistic construal of a statistical model $\mathcal{M}_\theta(\mathbf{z})$ so that its
specification relies solely on the statistical information contained in the data,
and at the same time nests the substantive model $\mathcal{M}_\psi(z)$ in its context via certain parameter restrictions $G(\theta, \varphi) = 0$. Viewing $\mathcal{M}_\theta(\mathbf{z})$ as a parametrization of
probabilistic structure of observable stochastic process $\{\mathbf{Z}_t, \ t \in \mathbb{N}\}$ underlying
the data Z_0, enables one to separate the validation facet (establishing the statistical adequacy of $\mathcal{M}_\theta(\mathbf{z})$) using thorough M-S testing and respecification, from
the inferential facet of posing substantive questions of interest. The substantive
and statistical information are reconciled by assessing the restrictions $G(\theta, \varphi) =$
0 in the context of a statistically adequate $\mathcal{M}_\theta(\mathbf{z})$. When the validity of these
restrictions is established, $\mathcal{M}_{\hat{\varphi}}(x)$ provides a sound basis for assessing the substantive adequacy of the underlying theory and learning from data about the
phenomenon of interest.

NOTES

Boumans and Hon, 'Introduction'

1. D. Diderot and J. d'Alembert (eds), *Encyclopédie ou Dictionnaire Raisonné des Sciences, des Arts et des Métiers*, 17 vols (Paris: Briasson, 1751–1765), vol. 5 (1755), entry: *Erreur*, p. 910.
2. A. de Morgan, *Formal Logic: or, The Calculus of Inference. Necessary and Probable* (London: Taylor and Walton, 1847), p. 237, italics in the original.
3. Z. Sng, *The Rhetoric of Error from Locke to Kleist* (Stanford, CA: Stanford University Press, 2010), pp. 3–4.
4. D. W. Bates, *Enlightenment Aberrations; Error and Revolution in France* (Ithaca, NY, and London: Cornell University Press, 2002), pp. 19–21.
5. G. Hon, 'Going Wrong: To Make a Mistake, to Fall into an Error', *Review of Metaphysics*, 49 (1995), pp. 3–20, on p. 6.
6. J. Derrida, *Memoirs of the Blind. The Self-Portrait and Other Ruins* (Chicago, IL: The University of Chicago Press, [1990] 1993), p. 13, italics in the original.
7. I. Newton, *The Mathematical Principles of Natural Philosophy*, trans. A. Motte, 3rd edn (New York: Prometheus Books, [1687/1726] 1995), p. 320, Rules of Reasoning in Philosophy, Rule III.
8. G. Galileo, *Dialogue Concerning the Two Chief World Systems*, trans. S. Drake, 2nd edn (Berkeley, CA: University of California Press, [1632] 1974), pp. 207–208.
9. F. Bacon, *The New Organon*, ed. L. Jardine and M. Silverthorne (Cambridge: Cambridge University Press, [1620] 2000), p. 10, preface.
10. Ibid., p. 13, Preface.
11. Ibid.
12. J. Bogen and J. Woodward, 'Saving the Phenomena', *Philosophical Review*, 97 (1988), pp. 303–52.
13. Ibid., p. 307.
14. Ibid., p. 312.
15. S. M. Stigler, *The History of Statistics: The Measurement of Uncertainty Before 1900* (Cambridge, MA: Harvard University Press, 1986).
16. L. Daston, 'The Moral Economy of Science', *Osiris*, 10 (1995), pp. 2–24.
17. Galileo, *Dialogue Concerning the Two Chief World Systems*; for an analysis see G. Hon, 'Kepler's Conception of Error in Optics and Astronomy: A Comparison with Galileo', in R. L. Kremer and J. Wlodarczyk (eds), *Studia Copernicana*, 42 (Warsaw, 2009), pp. 213–219.

18. A. Hald, 'Galileo's Statistical Analysis of Astronomical Observations', *International Statistical Review*, 54 (1986), pp. 211–20, on p. 212; see also A. Hald, *A History of Probability and Statistics and Their Applications Before 1750* (New York, 1990), pp. 149–60, and L. E. Maistrov, *Probability Theory: A Historical Sketch* (New York and London, 1974), pp. 32–4.

19. See Stigler, *The History of Statistics*, pp. 39–50.

20. Stigler, *The History of Statistics*.

21. Lippmann quoted in E. T. Whittaker and G. Robinson, *The Calculus of Observations* (London, 1924), p. 179.

22. Quoted in G. Hon, 'Towards a Typology of Experimental Errors: an Epistemological View', *Studies in History and Philosophy of Science*, 20 (1989), pp. 469–504, on p. 476.

23. Quoted in ibid, p. 476.

24. H. Margenau, *The Nature of Physical Reality, A Philosophy of Modern Physics* (New York: McGraw-Hill, 1950), p. 114.

25. Quoted in Hon, 'Towards a Typology of Experimental Errors', p. 477.

26. O. Morgenstern, *On the Accuracy of Economic Observations*, 2nd edn (Princeton, NJ: Princeton University Press, 1963), p. 13.

27. Hon, 'Towards a Typology of Experimental Errors', p. 475.

28. Quoted in Stigler, *The History of Statistics*, p. 309.

29. Ibid., italics in the original.

30. JCGM 104 (2009), *Evaluation of Measurement Data – An Introduction to the 'Guide to the Expression of Uncertainty in Measurement' and Related Documents*. Joint Committee for Guides in Metrology, p. 3, italics added.

31. JCGM 104, *Evaluation of Measurement Data*, p. 3.

32. I. J. Good, 'Degrees of Belief', *Encyclopedia of Statistical Sciences*, ed. S. Kotz and N. L. Johnson (New York: Wiley, 1982), vol. 2, p. 288.

33. Margenau, *The Nature of Physical Reality*, p. 162.

1 Karstens, 'The Lack of a Satisfactory Conceptualization of the Notion of Error in the Historiography of Science'

1. I thank the participants of the 'Error in the Sciences' conference in 2011 in Leiden, especially Eran Tal, for valuable comments which have led to a number of crucial adjustments in the text. The work is part of the research programme 'Philosophical Foundations of the Historiography of Science' financed by the Netherlands Organization for Scientific Research (NWO). I thank members of our HPS group at Leiden University for helpful discussions on the topics involved.

2. See G. Hon, 'Error: The Long Neglect, the One-Sided View and a Typology', in G. Hon, J. Schickore and F. Steinle (eds), *Going Amiss in Experimental Research* ([New York]: Springer 2009), p. 17: 'Error covers multiple sins. It is a multifarious epistemological phenomenon of great breadth and depth.'

3. Useful classifications in error types can be found in Hon, 'Towards a Typology of Experimental Errors', repeated in an updated version in Hon, 'Error: The Long Neglect, the One-Sided View and a Typology' and also in D. Allchin, 'Error Types', *Perspectives on Science*, 9 (2001), pp. 38–58.

4. Hon, 'Going Wrong', proposes reserving the term error for unforeseen ways of going wrong, while mistakes are the result of ways to go wrong that could have been foreseen and

thus also could have been avoided. This is a clarifying analysis whereas the distinction of K. Gavroglu, 'A Pioneer Who Never Got It Right: James Dewar and the Elusive Phenomena of Cold', in Hon, Schickore and Steinle (eds), *Going Amiss in Experimental Research*, pp. 137–57, between going amiss (unforeseen) and error (foreseen) appears unnecessarily complicating to me. I will not adopt technical usage of these terms since they are not generally acknowledged. The main thing is to keep in mind the importance of the distinction.

5. See G. Sarton, *The History of Science and the New Humanism* (New York: H. Holt and Company, 1931) and G. Sarton, *A Guide to the History of Science. A First Guide for the Study of the History of Science with Introductory Essays on Science and Tradition* (Waltham, MA: Chronica Botanica, 1952).

6. There are many points of overlap between Sarton's 'New Humanism' and Nietzsche's famous essay 'Vom Nutzen und Nachteil der Historie für das Leben'. Sarton's claims are also almost identical to C. P. Snow, *The Two Cultures and the Scientific Revolution* (Cambridge: Cambridge University Press, 1959). In *The Two Cultures* no reference to Sarton is made, however. For comments on the similarity between Sarton and Snow see also H. F. Cohen, 'De Wetenschapsrevolutie van de 17ᵉ eeuw en de Eenheid van het Wetenschappelijk Denken', in W. W. Mijnhardt and B. Theunissen (eds), *De Twee Culturen. De Eenheid van Kennis en haar Teloorgang* (Amsterdam: Rodopi, 1988), pp. 3–14.

7. Sarton, *A Guide to the History of Science*, pp. 41–2.

8. Ibid., p. 4.

9. A. Koyré, *Galileo Studies* (Sussex: Harvester Press, 1978), p. 66.

10. Another well-known historian of Koyré's generation, E. J. Dijksterhuis, valued the study of error for yet another reason. He argued that more than anything else, the historian of science should try to retrieve the process of scientific thinking. For this a focus on presumably flawless end products only would not be sufficient. Of much more interest could be the blot papers of scientists, not in order to find errors in the scribblings of 'great men', but to reconstruct the development of ideas before they became expressed in 'logically faultlessly ordered systems of definitions, axioms and hypotheses'. E. J. Dijksterhuis, *De Mechanisering van het Wereldbeeld* (Amsterdam: J. M. Meulenhof, 1950), p. 375.

11. Koyré, *Galileo Studies*, p. 109. The role that actual experiments played in Galileo's conceptual breakthrough has been a point of discussion in historiography. Koyré denied that Galileo actually performed experiments and claimed that he imagined ideal situations. For discussion on thought experiments see J. W. McAllister, 'The Evidential Significance of Thought Experiments in Science', *Studies in History and Philosophy of Science*, 27 (1996), pp. 233–50. M. Van Dyck, 'The Paradox of Conceptual Novelty and Galileo's Use of Experiments', *Philosophy of Science* (2005), pp. 864–75, argues that Galileo actually undertook explorative experiments and that these had an important evidential role for him.

12. Koyré, *Galileo Studies*, pp. 42–3.

13. Ibid., p. 66. Perhaps this view can be traced back to E. Mach, *Erkenntnis und Irrtum. Skizzen zur Psychologie der Forschung* (Darmstadt: Wissenschaftliche Buchgesellschaft, 1976), since it is identical to it.

14. Hon, 'Error: The Long Neglect, the One-Sided View and a Typology', p. 15.

15. Whiggish representations of past science are more likely to be found in popular science books or among scientists that at some point in their career started to write historical overviews of their field.

16. J. Schickore, 'Error as Historiographical Challenge: The Infamous Globule Hypothesis', in Hon, Schickore and Steinle (eds), *Going Amiss in Experimental Research*, pp. 27–45, on p. 31.

17. K. R. Popper, *Conjectures and Refutations: The Growth of Scientific Knowledge* (London: Routledge and Kegan Paul, 1963).

18. I borrow this observation from Hon, 'Error: The Long Neglect, the One-Sided View and a Typology', p. 11.

19. Ibid., pp. 12–13.

20. For the connection see especially K. R. Popper, 'Evolutionary Epistemology', in J. W. Pollard (ed.), *Evolutionary Theory. Paths into the Future* (London: John Wiley & Sons, 1984). Initially Popper was reluctant to interpret his theory as Darwinistic since he thought that Darwinism itself was not falsifiable and hence not a well formulated theory. The identification, then, would also cause Popper's own theory to be badly formulated. Later on he changed his mind about the tautological character of the survival of the fittest idea and maintained that it could be empirically tested.

21. I. Lakatos, 'History of Science and Its Rational Reconstructions', in *PSA: Proceedings of the Biennial Meeting of the Philosophy of Science Association* (1970), pp. 91–136.

22. Lakatos, 'History of Science and Its Rational Reconstructions', p. 118.

23. J. Schickore, '"Through Thousands of Errors We Reach the Truth" – But How? On the Epistemic Roles of Error in Scientific Practice', *Studies in the History and Philosophy of Science*, 36 (2005), pp. 539–56, on p. 541, makes this observation too.

24. L. Laudan, *Progress and Its Problems. Toward a Theory of Scientific Growth* (Berkeley, CA, [1977]), p. 128.

25. Ibid., p. 127.

26. Note that Laudan also grants a role to the comparative evaluation of other theoretical virtues such as simplicity, predictive success etc. It is, however, problem-solving capacity that must be seen as the overriding factor governing theory choice.

27. L. Laudan, *Beyond Positivism and Relativism: Theory, Method and Evidence* (Oxford: Westview Press, 1996), p. 77.

28. Laudan, *Progress and Its Problems*, pp. 221–2.

29. D. G. Mayo, *Error and the Growth of Experimental Knowledge* (Chicago, IL, and London: University of Chicago Press, 1996), p. xii.

30. Ibid., pp. 184–5. Note that complete exclusion of error is not possible, hence error control is the main aim of her project.

31. Quoted from M. Boon, 'Instruments in Science and Technology', in J.-K. Berg Olsen, S. A. Pedersen and V. F. Hendricks (eds), *A Companion to Philosophy of Technology* (Oxford: Wiley-Blackwell, 2009), pp. 78–84.

32. A. F. Chalmers, *What is This Thing Called Science?* 3rd edn (Berkshire: Open University Press, 1999), p. 202.

33. Mayo, *Error and the Growth of Experimental Knowledge*, p. x.

34. Ibid., p. 15, emphasis in original text.

35. Allchin, 'Error Types'. Mayo's treatment of the subject does not show much interest in the history of science. It is also by and large a story of prospective error. We will come back to these points in the analysis on pp. 30–1 of the main chapter text.

36. Mayo, *Error and the Growth of Experimental Knowledge*, p. 150. Mayo refers to Laudan extensively throughout her book.

37. Mayo, *Error and the Growth of Experimental Knowledge*, p. 55.

38. R. K. Merton, *The Sociology of Science: Theoretical and Empirical Investigations* (Chicago, IL: The University of Chicago Press, 1973), p. 11.

39. D. Bloor, *Knowledge and Social Imagery* (Chicago, IL, and London: University of Chicago Press, 1991 [1976]), p. 8. Bloor actually attributes this to Laudan, which is a gross

oversimplification of the latter's position. Laudan does not speak about truth anymore and it is also not the case that 'nothing' makes people do things that are correct, since all the cognitive relevant factors should explain this. Bloor, however, attacks the self-explanatory character of such explanations. See also the debate between Laudan and Bloor in *Philosophy of the Social Sciences*: L. Laudan, 'The Pseudo Science of Science', *Philosophy of the Social Sciences*, 11 (1981), pp. 173–98, and D. Bloor, 'The Strengths of the Strong Programme', *Philosophy of the Social Sciences*, 11 (1981), pp. 199–213.

40. Bloor, *Knowledge and Social Imagery*, p. 1. The fact that scientists create the idea that such specialties do exist is thought to be part of their self-fashioning and shaping of their professional identity. M. Mulkay and G. N. Gilbert, 'Accounting for Error: How Scientists Construct Their Social World When They Account for Correct and Incorrect Belief', *Sociology*, 16 (1982), pp. 165–83, aims to demonstrate this.

41. Bloor, *Knowledge and Social Imagery*.

42. Schickore, '"Through Thousands of Errors We Reach the Truth" – But How?'.

43. A key example is S. Schaffer and S. Shapin, *Leviathan and the Air-pump: Hobbes, Boyle and the Experimental Life* (Princeton, NJ: Princeton University Press, 1985). See J. Golinski, *Making Natural Knowledge: Constructivism and the History of Science* (Chicago, IL, and London: Chicago University Press, 1998) for a survey of the historiography of science after the social turn and the further development of social constructivism in later years.

44. See for example H. Collins, *Changing Order: Replication and Induction in Scientific Practice* (Beverley Hills, CA, and London: Sage, 1985) or H. Petroski, *Success through Failure: The Paradox of Design* (Princeton, NJ: Princeton University Press, 2006).

45. This is the way P. Galison, 'Author of Error', *Social Research*, 72:1 (2005), pp. 63–76, studies a curious episode in the history of science in which Enrico Fermi mistakenly received the Nobel Prize in Physics for the discovery of nuclear fusion, which he did *not* discover, as it later turned out.

46. These extensions are based on extensions of the symmetry principle. For an overview of these see D. Pels, 'The Politics of Symmetry', *Social Studies of Science*, 26:2 (1996), pp. 277–304.

47. Collins, *Changing Order* is an earlier presentation of such a network model.

48. J. Secord, 'Knowledge in Transit', *Isis*, 95 (2004), pp. 654–72, can be read as the inaugural lecture of this 'new' research programme for historiography of science.

49. See P. Smith and B. Schmidt (eds), *Making Knowledge in Early Modern Europe: Practices, Objects, and Texts, 1400–1800* (Chicago, IL: Chicago University Press, 2007).

50. Note that this is a completely different trial and error procedure than Popper's falsification procedure, as the latter does not inquire into the process of actual experimentation.

51. I first came across the notion of the fertile error in the issue of *Social Research*, 72:1 (2005), which was entirely devoted to the study of error. The first to use the term was probably W. C. Wimsatt, 'False Models as Means to Truer Theories', in M. H. Nitecki and A. Hoffmann (eds), *Neutral Models in Biology* (New York: Oxford University Press, 1987), pp. 23–55.

52. See H. C. Ohanian, *Einstein's Mistakes: The Human Failings of Genius* (New York and London: W. W. Norton & Company, 2008).

53. See for example E. Jorink, *'Geef Zicht aan de Blinden' Constantijn Huygens, René Descartes en het Boek der Natuur* (Leiden: Primavera Pers, 2008).

54. Hon, 'Going Wrong'.

55. Wimsatt, 'False Models as Means to Truer Theories'.

56. Bad wrong may still be better than confusion. As Wolfgang Pauli once exclaimed when he was confronted with the paper of a young and presumably promising physicist: 'It is not even wrong!', in other words: 'I can do nothing with it'.

57. Some, for example Laudan's problem-solving approach, are context-sensitive but maintain an asymmetrical analysis of past science. This requires a definition of rationality granting it a privileged status and this is the root of the problems we have with the 'errors as obstacles' approach.

58. Hon, 'Error: The Long Neglect, the One-Sided View and a Typology', pp. 16–20, has argued that because of this Mayo has not delivered a satisfactory theory of experiment. This holds for the design of experiments, the practical difficulties as well as for the possible modes of interpretations of experimental results. The narrow treatment may be explained by the fact that Mayo is fighting a battle against the theory oriented philosophies of science and, in comparison to those approaches, is already making a huge case for more sensitivity towards experiments. However, when a historical view is opened up it appears that an even richer treatment of this subject is in store.

59. This is done in terms of strategies of research and specifying typical research situations. See Staley (this volume) and L. Darden, *Theory Change in Science: Strategies from Mendelian Genetics* (New York and Oxford: Oxford University Press, 1991).

60. This criticism has been levelled against Mayo by M. Weber, 'The Crux of Crucial Experiments: Duhem's Problems and Inference to the Best Explanation', *British Journal Philosophy of Science*, 60 (2009), pp. 19–49, especially pp. 31–3.

61. H. Chang, 'Review: Deborah Mayo, Error and the Growth of Experimental Knowledge', *British Journal for the Philosophy of Science*, 48 (1997), pp. 455–9.

62. Mayo, *Error and the Growth of Experimental Knowledge*, p. 150.

63. See also K. Nickelsen and G. Grasshoff, 'Concepts from the Bench: Hans Krebs, Kurt Henseleit and the Urea Cycle', in Hon, Schickore and Steinle (eds), *Going Amiss in Experimental Research*, pp. 91–118, on p. 93, for this argument.

64. N. Jardine, 'Whigs and Stories: Herbert Butterfield and the Historiography of Science', *History of Science*, 41 (2003), pp. 125–40.

65. This argument is made in O. Kiss, 'Meaningful Mistakes', in M. Fehér, O. Kiss and L. Ropolyi (eds), *Hermeneutics and Science* (Dordrecht: Kluwer Academic Publishers, 1999), pp. 125–33. Most of the arguments against Whig history can be found in J. Schuster, 'The Problem of Whig History in the History of Science', ch. 3 in *The Scientific Revolution: An Introduction to the History & Philosophy of Science*, available at descartes-agonistes.com.

66. Elzinga, 'Wetenschapsgeschiedenis van overzee: Congresverslag', *Skript*, 8 (1987), pp. 263–70, on p. 267.

67. Hon, Schickore and Steinle (eds), *Going Amiss in Experimental Research*, introduction.

68. R. Sargent, *The Diffident Naturalist. Robert Boyle and the Philosophy of Experiment* (Chicago, IL, and London: University of Chicago Press, 1995), p. 20.

69. C. Ray, 'The Cosmological Constant: Einstein's Greatest Mistake?', *Studies in History and Philosophy of Science*, 21 (1990), pp. 589–604; Ohanian, *Einstein's Mistakes*. Ray shows that the mistake may not have been so bad after all. The cosmic constant addresses one of the persistent problems in physics: how to explain the stability of the universe, and is still in use today.

70. M. Weber, 'Wissenschaft als Beruf', in D. Kaesler (ed.), *Max Weber: Schriften 1894–1922* (Stuttgart: Alfred Kröner, 2002), pp. 486–7. [Translation: Every scientist knows

that what he has worked at will be outdated in 10, 20, 50 years ... every scientific 'fulfilment' triggers new 'questions' and begs to be surpassed and come of age.]

71. G. Hon, 'Putting Error to (Historical) Work: Error as a Tell-Tale in the Studies of Kepler and Galileo', *Centaurus*, 46 (2004), pp. 58–81. A broad account of changes in mentality in the sixteenth to nineteenth centuries is the subject of L. Daston, 'Scientific Error and the Ethos of Belief', *Social Research*, 72:1 (2005), pp. 1–28.

72. S. Weinberg, 'Sokal's Hoax', *New York Review of Books*, 43:13 (1996), pp. 11–15, and J. Frercks, 'Going Right and Making it Wrong: The Reception of Fizeau's Ether-Drift Experiment of 1859', in Hon, Schickore and Steinle (eds), *Going Amiss in Experimental Research*, pp. 179–210, give examples of strange experimental results obtained by Thompson (Kelvin) and Fizeau that can only be made sense of if present-day knowledge about the phenomena they were testing is included in the historical analysis. Incidentally, the very possibility of the restaging of experiments as in the work of Pamela Smith is also based on the assumption of continuity of phenomena.

73. K. Alder, 'The History of Science, or, an Oxymoronic Theory of Relativistic Objectivity', in L. Kramer and S. Maza (eds), *A Companion to Western Historical Thought* (London 2002), pp. 297–318, also makes the point that historiography of science lost its status as an intensifier of the epistemic debate after the 'social' turn.

74. Only Pickering is not bothered by this. He sees it as an asset that his mangle idea may turn out to be a theory of everything. See A. Pickering and K. Guzik (eds), *The Mangle in Practice: Science, Society and Becoming* (Durham, NC: Duke University Press, 2008).

75. See also B. Karstens, 'Towards a Classification of Approaches to the History of Science', *Organon*, 43 (2011), pp. 47–52, for a related analysis of approaches to past science.

76. H. Nowotny, P. Scott and M. Gibbons, *Re-Thinking Science: Knowledge and the Public in an Age of Uncertainty* (Oxford: Polity, 2011).

77. The concept of 'going amiss' is introduced in Hon, Schickore and Steinle (eds), *Going Amiss in Experimental Research*. Most of the case studies in this volume are highly relevant to further developing a novel way of dealing with errors in past science.

78. H. Chang, 'We Have Never Been "Whiggish" (About Phlogiston)', *Centaurus*, 51 (2009), pp. 455–9, on p. 254.

79. But not vicious anachronism; for the distinction see N. Jardine, 'Uses and Abuses of Anachronism in the History of the Sciences', *History of Science*, 38 (2000), pp. 251–70.

80. See B. Karstens, 'Bopp the Builder. Discipline Formation as Hybridization: The Case of Comparative Linguistics', in R. Bod, J. Maat and T. Westeijn (eds), *The Making of the Humanities. Volume II: From Early Modern to Modern Disciplines* (Amsterdam: Amsterdam University Press, 2012), pp. 103–27.

81. Laudan, *Beyond Positivism and Relativism*.

82. B. Latour, *Science in Action: How to Follow Scientists and Engineers through Society* (Cambridge, MA: Harvard University Press, 1987), for example, speaks of a ban on cognitive factors in the explanation of past science. The role of cognitive factors is, according to Latour, always secondary to other factors as it is seen as a derivative of these.

83. See J. Kagan, *Surprise, Uncertainty and Mental Structures* (Cambridge, MA: Harvard University Press, 2002) and especially G. Gigerenzer, *Rationality for Mortals: How People Cope with Uncertainty* (New York and Oxford: Oxford University Press, 2008).

84. See also G. Gigerenzer, 'I Think, Therefore I Err', *Social Research*, 72:1 (2005), pp. 195–218, reprinted in Gigerenzer, *Rationality for Mortals*, specifically on error.

85. See also P. Roth, 'Mistakes', *Synthese*, 136 (2003), pp. 389–408.

2 Staley, 'Experimental Knowledge in the Face of Theoretical Error'

1. P. Galison, *How Experiments End* (Chicago, IL: University of Chicago Press, 1987).
2. A. Franklin, *The Neglect of Experiment* (New York: Cambridge University Press, 1986); I. Hacking, *Representing and Intervening: Introductory Topics in the Philosophy of Natural Science* (Cambridge: Cambridge University Press, 1983); Mayo, *Error and the Growth of Experimental Knowledge*.
3. Franklin, *The Neglect of Experiment*; Mayo, *Error and the Growth of Experimental Knowledge*; P. Galison, *Image and Logic: A Material Culture of Microphysics* (Chicago, IL: University of Chicago Press, 1997).
4. A. F. Chalmers, 'Can Scientific Theories Be Warranted?', in D. G. Mayo and A. Spanos (eds), *Error and Inference: Recent Exchanges on Experimental Reasoning, Reliability, Objectivity, and Rationality* (New York: Cambridge University Press, 2009), pp. 58–72; M. Heidelberger, 'Theory-Ladenness and Scientific Instruments in Experimentation', in H. Radder (ed.), *The Philosophy of Scientific Experimentation* (Pittsburgh, PA: University of Pittsburgh Press, 2003), pp. 138–51; H. Radder, 'Technology and Theory in Experimental Science', in Radder (ed.), *The Philosophy of Scientific Experimentation*, pp. 152–73; K. W. Staley, 'Error-Statistical Elimination of Alternative Hypotheses', *Synthese*, 163 (2008), pp. 397–408.
5. Whether, in a given case, the approximate satisfaction of an assumption makes that assumption safe to use for purposes of inference is not easily decided. The assumption that approximately satisfied premises lead to approximately correct conclusions is a major source of error in statistical inference and is the motivation for the development of both robust statistics, see F. R. Hampel, E. M. Ronchetti, P. J. Rousseeuw and W. A. Stahel, *Robust Statistics: The Approach Based on Influence Functions* (New York: John Wiley, 1986) and P. Huber, *Robust Statistics* (New York: John Wiley, 1981), and mis-specification testing, see A. Spanos, 'Foundational Issues in Statistical Modeling: Statistical Model Specification and Validation', *Rationality, Markets and Morals*, 2 (2011), pp. 146–78; and see K. W. Staley, 'Strategies for Securing Evidence Through Model Criticism', *European Journal for Philosophy of Science*, 2 (2012), pp. 21–43, for a discussion.
6. Spanos, *Probability Theory and Statistical Inference*; Spanos, 'Foundational Issues in Statistical Modeling'; Staley 'Strategies for Securing Evidence Through Model Criticism'.
7. D. Chalmers, 'The Nature of Epistemic Space', in A. Egan and B. Weatherson (eds), *Epistemic Modality* (Oxford: Oxford University Press, 2011); K. DeRose, 'Epistemic Possibilities', *Philosophical Review*, 100 (1991), pp. 581–605; J. Hintikka, *Knowledge and Belief: An Introduction to the Logic of the Two Notions* (Ithaca, NY: Cornell University Press, 1962); A. Kratzer, 'What "Must" and "Can" Must and Can Mean', *Linguistics and Philosophy*, 1 (1977), pp. 337–55; J. MacFarlane, 'Epistemic Modals are Assessment-Sensitive', in A. Egan and B. Weatherson (eds), *Epistemic Modality* (Oxford: Oxford University Press, 2011); J. Salerno, Epistemic Modals: Relativism and Ignorance. Talk given at Midwest Epistemology Workshop, Saint Louis University, September 2009.
8. Occurrences of the word 'possible' and its cognates should henceforth be understood epistemically, unless otherwise specified.
9. It should not be assumed that what is epistemically possible for an epistemic agent will always be known to be epistemically possible for that agent. Arguably, nothing that Isaac

Newton knew, and no information to which he had access, ruled out the possibility that gravity is a manifestation of the curvature of space. There is a sense of possibility, distinct from that here invoked, in which curved space became possible much later. To my knowledge, the semantics of this latter notion have not been seriously tackled by philosophers. For purposes of the present discussion, I will assume that *discovering* a new possibility does not *create* a new possibility.

10. See e.g. J. Stegenga, 'Robustness, Discordance, and Relevance', *Philosophy of Science*, 76:5 (2009), pp. 650–61; J. Woodward, 'Some Varieties of Robustness', *Journal of Economic Methodology*, 13 (2006), pp. 219–40; and K. Coko and J. Schickore, *Robustness and its Kith and Kin: A Meta-Analysis* (2011) for a survey.

11. K. W. Staley, 'Robust Evidence and Secure Evidence Claims', *Philosophy of Science*, 71 (2004), pp. 467–88.

12. I. Newton, *The Principia: Mathematical Principles of Natural Philosophy*, ed. and trans. I. B. Cohen and A. Whitman (Berkeley, CA: University of California Press, 1999), pp. 802–11.

13. W. Harper, 'Newton's Argument for Universal Gravitation', in I. B. Cohen and G. E. Smith (eds), *The Cambridge Companion to Newton* (New York: Cambridge University Press, 2002), pp. 174–201; W. Harper, *Isaac Newton's Scientific Method: Turning Data into Evidence about Gravity and Cosmology* (New York: Oxford University Press, 2011).

14. Harper, 'Newton's Argument for Universal Gravitation', pp. 175–7.

15. Newton, *The Principia*, pp. 539–45; Harper, 'Newton's Argument for Universal Gravitation', pp. 180–1.

16. Newton, *The Principia*, p. 417.

17. Letter to Newton, 18 March 1713, quoted in Harper, *Isaac Newton's Scientific Method*.

18. H. Stein, '"From the Phenomena of Motions to the Forces of Nature": Hypothesis or Deduction?', *PSA: Proceedings of the Biennial Meeting of the Philosophy of Science Association, 1990* (1990), pp. 209–22.

19. Harper, *Isaac Newton's Scientific Method*.

20. Newton, *The Principia*, pp. 427–8.

21. Ibid., p. 428.

22. Harper, *Isaac Newton's Scientific Method*.

23. Ibid., ch. 9.IV.2.

24. Newton, *The Principia*, p. 943.

25. I. Newton, *Philosophical Writings*, ed. A. Janiak (Cambridge: Cambridge University Press, 2004), p. 102; A. Janiak, *Newton as Philosopher* (Cambridge: Cambridge University Press, 2010).

26. W. Harper, 'Newton's Methodology and Mercury's Perihelion Before and After Einstein', *Philosophy of Science*, 74:5 (2007), pp. 932–42.

27. C. M. Will, *Theory and Experiment in Gravitational Physics*, 2nd edn (New York: Cambridge University Press, 1993), p. 22.

28. C. M. Will, 'The Confrontation between General Relativity and Experiment', *Living Reviews in Relativity*, 9:3 (2006), p. 29.

29. D. G. Mayo, 'Theory Testing, Statistical Methodology, and the Growth of Experimental Knowledge', in P. Gärdenfors, J. Wolinski and K. Kijania-Placek (eds), *In the Scope of Logic, Methodology, and Philosophy of Science* (Dordrecht: Kluwer, 2002), pp. 171–90; D. G. Mayo, 'Learning from Error, Severe Testing, and the Growth of Theoretical Knowledge', in Mayo and Spanos (eds), *Error and Inference*, pp. 28–57.

30. I. I. Shapiro, 'New Method for the Detection of Light Deflection by Solar Gravity', *Science*, 157:3790 (1967), pp. 806–8.

31. S. S. Shapiro, J. L. Davis, D. E. Lebach and J. S. Gregory, 'Measurement of the Solar Gravitational Deflection of Radio Waves Using Geodetic Very-Long-Baseline Interferometry Data, 1979–1999', *Physical Review Letters*, 92 (2004), 121101.

32. Ibid.

33. Will, *Theory and Experiment in Gravitational Physics*, p. 126.

34. I note here how these principles might be stated: *WEP*: if an uncharged test body is placed at an initial event in spacetime and given an initial velocity there, then its subsequent trajectory will be independent of its internal structure and composition. *LLI*: the outcome of any local nongravitational test experiment is independent of the velocity of the experimental apparatus. *LPI*: the outcome of any local nongravitational test experiment is independent of its spacetime location. See Will, *Theory and Experiment in Gravitational Physics*, p. 22.

35. Ibid., p. 207.

36. Ibid., p. 10.

37. Will, 'The Confrontation between General Relativity and Experiment', p. 26.

38. A. Lightman and D. Lee, 'Restricted Proof that the Weak Equivalence Principle Implies the Einstein Equivalence Principle', *Physical Review D*, 8 (1973), pp. 364–76. See D. Mattingly, 'Modern Tests of Lorentz Invariance', *Living Reviews in Relativity*, 8 (2005) for a review of this and other parametric frameworks as applied to tests of Lorentz invariance.

39. The restriction, more specifically, is to theories that describe the center-of-mass acceleration of an electromagnetic test body in a static, spherically symmetric gravitational field, such that the dynamics for particle motion is derivable from a Lagrangian. The parameters T and H appear in the Lagrangian; ε and μ appear in the 'gravitationally modified Maxwell equations' (GMM). Lightman and Lee, 'Restricted Proof that the Weak Equivalence Principle Implies the Einstein Equivalence Principle', argue that 'all theories we know of' have GMM equations of the type needed, and that all but one theory (which they treat separately) can be represented in terms of the appropriate Lagrangian.

40. Will, *Theory and Experiment in Gravitational Physics*, p. 31.

41. Ibid., p. 62.

42. D. Colladay and V. A. Kostelecký, 'Lorentz-Violating Extension of the Standard Model', *Physical Review D*, 58 (1998), 116002; V. A. Kostelecký, 'Gravity, Lorentz Violation, and the Standard Model', *Physical Review D*, 69 (2004), 105009.

43. D. Bear, R. Stoner, R. Walsworth, V. Kostelecký and C. Lane, 'Limit on Lorentz and CPT Violation of the Neutron Using a Two-Species Noble-Gas Maser', *Physical Review Letters*, 85 (2000), pp. 5038–41; D. Bear, R. Stoner, R. Walsworth, V. Kostelecký and C. Lane, 'Erratum: Limit on Lorentz and CPT Violation of the Neutron Using a Two-Species Noble-Gas Maser', *Physical Review Letters*, 89 (2002), 209902.

44. Mattingly, 'Modern Tests of Lorentz Invariance'; Will, 'The Confrontation between General Relativity and Experiment'.

45. DeRose, 'Epistemic Possibilities'.

46. Harper, 'Newton's Methodology and Mercury's Perihelion Before and After Einstein'; Harper, *Isaac Newton's Scientific Method*; Newton, *The Principia*.

47. Mayo, *Error and the Growth of Experimental Knowledge*.

48. Ibid., ch. 6.

49. Staley, 'Error-Statistical Elimination of Alternative Hypotheses'.

50. Egan and Weatherson (eds), *Epistemic Modality*.

3 Mayo, 'Learning from Error: How Experiment Gets a Life (of its Own)

1. Hacking, *Representing and Intervening*; I. Hacking, 'The Self-Vindication of the Laboratory Sciences', in A. Pickering (ed.), *Science as Practice and Culture* (Chicago, IL: University of Chicago Press, 1992), pp. 29–64; I. Hacking, 'Statistical Language, Statistical Truth, and Statistical Reason: The Self-Authentication of a Style of Scientific Reasoning', in E. McMullin (ed.), *The Social Dimensions of Science* (Notre Dame, IN: University of Notre Dame Press, 1992), pp. 130–57.

2. If there is a need to restore the more usual distinction between experimental and observational research, the former might be dubbed 'manipulative experiment' and the latter 'observational experiment'.

3. In Mayo, *Error and the Growth of Experimental Knowledge*, errors number 4 and 5 were collapsed under the rubric of 'experimental assumptions'.

4. I discuss this at length elsewhere, e.g. D. G. Mayo, 'Novel Evidence and Severe Tests', *Philosophy of Science*, 58 (1991), pp. 523–52. Reprinted in *Philosopher's Annual*, 14 (Atascadero, CA: Ridgeview Publishing Co., 1991), pp. 203–32; Mayo, *Error and the Growth of Experimental Knowledge*; D. G. Mayo, 'How to Discount Double-Counting When It Counts: Some Clarifications', *British Journal of Philosophy of Science*, 59 (2008), pp. 857–79; D. G. Mayo and M. Kruse, 'Principles of Inference and Their Consequences', in D. Cornfield and J. Williamson (eds), *Foundations of Bayesianism* (Dordrecht: Kluwer Academic Publishers, 2001), pp. 381–403; D. G. Mayo and D. Cox, 'Frequentist Statistics as a Theory of Inductive Inference', in J. Rojo (ed.), *Optimality: The Second Erich L. Lehmann Symposium*, vol. 49, Lecture Notes-Monograph Series (Institute of Mathematical Statistics, 2006), pp. 77–97. Reprinted in D. G. Mayo and A. Spanos (eds) *Error and Inference: Recent Exchanges on Experimental Reasoning, Reliability, and the Objectivity and Rationality of Science* (Cambridge: Cambridge University Press, 2010), pp. 247–75.

5. D. C. Gajdusek and V. Zigas, 'Degenerative Disease of the Central Nervous System in New Guinea; the Endemic Occurrence of Kuru in the Native Population', *New England Journal of Medicine*, 257:20 (1957), pp. 974–8. This work won Gajdusek a Nobel Prize in 1976.

6. The transmissibility of TSEs was accidentally demonstrated in 1937, when a population of Scottish sheep was inoculated against a common virus with a formalin extract of brain tissue unknowingly derived from an animal with scrapie. After two years, nearly 10 per cent of the flock developed scrapie. Scrapie was subsequently transmitted experimentally to sheep.

7. S. B. Prusiner, 'Novel Proteinaceous Infectious Particles Cause Scrapie', *Science*, 216:4542 (1982), pp. 136–44.

8. T. S. Kuhn, *The Structure of Scientific Revolutions* (Chicago, IL: University of Chicago Press, 1962), p. 122.

9. Ibid., p. 7.

10. Actually, the very constraints that Kuhn accords to 'normal science' or testing within a paradigm are at the heart of what gives stability to experimental effects. As I argue in Chapter 2 of Mayo, *Error and the Growth of Experimental Knowledge* (some may consider this a radical 'deconstruction' of Kuhn), normal science can be seen to describe the highly constrained, local experimental probing. Taking the stringent demands of normal science seriously forces the experimenter to face the music, to admit he has not solved his problem rather than seek to conveniently change the problem. Where Kuhn errs is in

throwing away the stringent demands he recognizes for probing within a paradigm when it comes to theory change. Moreover, Kuhn gives no argument for postulating a radical discontinuity between the role of evidence in conducting normal tests, and in finding flaws in a large paradigm ('revolutionary' science).

11. There is a corresponding argument that H is an incorrect construal of data. These arguments from error can equivalently be put in terms of inferring H (or not-H) with severity.

12. C. S. Peirce, *Collected Papers*, 6 vols, ed. C. Hartshorne and P. Weiss (Cambridge, MA: Harvard University Press, 1931–35), vol. 7, p. 1115 (ed. note).

13. Ibid., vol. 7, p. 59.

14. See Mayo, *Error and the Growth of Experimental Knowledge*; D. G. Mayo and A. Spanos, 'Error Statistics', in P. S. Bandyopadhyay and M. R. Forster (eds), *Philosophy of Statistics*, *Handbook of the Philosophy of Science* (Oxford: Elsevier, 2011), pp. 151–196.

15. P. Achinstein, 'Why Philosophical Theories of Evidence Are (and Ought to Be) Ignored by Scientists', *Philosophy of Science*, 67 (2000), pp. S180–S192. Symposia Proceedings, edited by D. Howard.

16. Popper once wrote to me that he regretted not having learned modern statistical methods. If he had, he might have seen error probabilities as offering the third way he sought. That was clearly Peirce's way, although he anticipated modern statistics.

17. A. Musgrave, 'Logical versus Historical Theories of Confirmation', *BJPS*, 25 (1974), pp. 1–23.

18. I am at odds here with others, and for different reasons. See for example L. Laudan, 'Anomaly of Affirmative Defenses', in Mayo and Spanos (eds) *Error and Inference* (2010), pp. 376–96; D. G. Mayo, 'Error and the Law, Exchanges with Larry Laudan', in Mayo and Spanos (eds) *Error and Inference* (2010), pp. 397–409.

19. Mayo, *Error and the Growth of Experimental Knowledge*.

20. A. Musgrave, 'Critical Rationalism, Explanation, and Severe Tests', in Mayo and Spanos (eds), *Error and Inference* (2010), pp. 81–112, on p. 108. While these are Musgrave's words, he claims to be summarizing the views of Chalmers, 'Can Scientific Theories Be Warranted?' (ibid., pp. 58–72) and L. Laudan, 'How About Bust? Factoring Explanatory Power Back into Theory Evaluation', *Philosophy of Science*, 64 (1997), pp. 303–16.

21. D. G. Mayo, 'Toward Progressive Critical Rationalism, Exchanges with Alan Musgrave', in Mayo and Spanos (eds) *Error and Inference* (2010), pp. 115–24.

22. W. Salmon, 'The Appraisal of Theories: Kuhn Meets Bayes', in A. Fine, M. Forbes and L. Wessels (eds), *PSA 1990*, vol. 2 (East Lansing, MI: Philosophy of Science Association, 1991), pp. 325–32, on p. 329.

23. Viewing evidential appraisal as unable to live a life apart from cost benefit analysis is a conception I reject. It has done much damage in evidence-based policy: if disagreements about interpreting data are indistinguishable from disagreements about matters of subjective opinions and values, then each side gets its own scientific experts, often resulting in 'junk science' all around. We can ask: should data on the frequency of BSE/mad cow disease be interpreted according to the economic values of the beef industry? Probably not, which is why obtaining sound inferences needs to be distinct from, and not inextricably intertwined with, subsequent policy decisions.

24. Hacking, *Representing and Intervening*.

25. One can depict the test in myriad ways. The artefact error might be seen as standing in for a null hypothesis; it asserts the presence of a given error.

26. But it was also the basis for constructing possible alternatives, such as virions.
27. S. B. Prusiner, 'Kuru with Incubation Periods Exceeding Two Decades', *Annuals of Neurology*, 12:1 (1982.), pp. 1–9; S. B. Prusiner, 'Human Prion Diseases and Neuro-degeneration', in S. B. Prusiner (ed.), *Prions, Prions, Prions*, vol. 207, Current Topics in Microbiology and Immunology (Berlin: Springer-Verlag, 1996), pp. 1–19; S. B. Prusiner, 'Prion Diseases and the BSE Crisis', *Science*, 278:5336 (1997), pp. 245–51.
28. Magic angle spinning was invented by physicist E. Raymond Andrew. For a discussion, see http://www.magnet.fsu.edu/education/tutorials/tools/probes/magicangle.html. [accessed 19 October 2013].
29. Nowadays, researchers know how to perform what is called cyclical amplification: lopping off the ends of the pathogenic (beta) helix, they get huge amounts of PrP-Sc starting with minute quantities. This affirms a correct understanding of the misfolding mechanism (pathogenic prions propagate through misfolding at the ends of the helix) and also gives an important new tool to detect prion disease in living animals for the first time.
30. Hacking, *Representing and Intervening*.
31. Franklin, *The Neglect of Experiment*.
32. I sometimes call these arguments from conspiracy, rigging or gellerization, e.g. Mayo, *Error and the Growth of Experimental Knowledge*; D. G. Mayo, 'Learning from Error: The Theoretical Significance of Experimental Knowledge', *Modern Schoolman*, 87: 3/4 (2010) (guest ed. K. Staley), *Experimental and Theoretical Knowledge, The Ninth Henle Conference in the History of Philosophy*, pp. 191–217.

4 Mari and Giordani, 'Modelling Measurement: Error and Uncertainty'

1. One of the authors is a member of the Joint Committee on Guides in Metrology (JCGM) Working Group 2 (VIM). The opinion expressed in this chapter does not necessarily represent the view of this working group.
2. As an example, consider the following quotation from a well-known textbook by J. R. Taylor, *An Introduction to Error Analysis – The Study of Uncertainties in Physical Measurements* (Sausalito, CA: University Science Books, 1997): 'All measurements, however careful and scientific, are subject to some uncertainties. Error analysis is the study and evaluation of these uncertainties … The analysis of uncertainties, or "errors," is a vital part of any scientific experiment'. 'Error' and 'uncertainty' seem to be used here interchangeably.
3. Quoted from the *International Vocabulary of Metrology*, 3rd edn, 'VIM3' henceforth JCGM 200, *International Vocabulary of Metrology – Basic and General Concepts and Associated Terms (VIM)*, 3rd edn (2008 version with minor corrections) (Joint Committee for Guides in Metrology, 2012), at http://www.bipm.org/en/publications/guides/vim.html. [accessed 4 October 2013].
4. Henceforth JCGM 100, *Evaluation of Measurement Data – Guide to the Expression of Uncertainty in Measurement* (GUM, originally published in 1993) (Joint Committee for Guides in Metrology, 2008), at http://www.bipm.org/en/publications/guides/gum.html. [accessed 4 October 2013].
5. Whether the measurable entities are quantities or, more generically, properties, is an issue for measurement science, but it is immaterial here. Hence we will maintain the traditional, restricted, version in our discussion, which may be generalized without difficulty to the case of ordinal and classificatory (or 'nominal') entities. It should also be noted that indi-

vidual quantities are instances of general quantities (e.g. the length if this table *is a* length). This justifies the usual loose terminological habit of leaving the distinction implicit.

6.   The realist standpoint rests on a tradition, dating back to the Pythagorean philosophers and championed by scientists such as Galilei and Kepler, which conceives of the world as having a definite mathematical structure: the world 'is written in mathematical characters' (Galilei) since 'numbers are in the world' (Kepler).

7.   The instrumentalist standpoint rests on a tradition, dating back to the Protagorean philosophers and championed by philosophers such as Carnap, which conceives of the world as being known in virtue of the adoption of a conceptual framework: 'We introduce the numerical concept of weight by setting up a procedure for measuring it. It is *we* who assign numbers to nature. The phenomena themselves exhibit only qualities we observe. Everything numerical ... is brought in by ourselves when we devise procedures for measurement', R. Carnap, *Philosophical Foundations of Physics* (New York: Basic Books, 1966), p. 100. In addition, Carnap construes quantitative concepts as essential tools for introducing quantitative laws and views such laws as essential tools for making previsions: 'The most important advantage of the quantitative law ... is not its brevity, but rather the use that can be made of it. Once we have the law in numerical form, we can employ that powerful part of deductive logic we call mathematics and, in that way, make predictions'. This conception of the role of quantitative concepts and laws justifies the use of the term 'instrumentalism' with respect to the proposed standpoint. It is also worth noting that such a standpoint is based on operationalist assumptions – see P. W. Bridgman, *The Logic of Modern Physics* (New York: Macmillan, 1927) – about the definition of scientific concepts. It is typically adopted by scholars influenced by the view that 'pure data does not exist' – see T. S. Kuhn, *The Structure of Scientific Revolutions* (Chicago, IL: University of Chicago Press, 1970), because 'data are always theory laden' – N. Hanson, *Patterns of Discovery* (Cambridge: Cambridge University Press, 1958).

8.   JCGM 200, *International Vocabulary of Metrology*.

9.   In this respect the 'Uncertainty Approach' is rather delicate, since the contemporary presence of different truths on the same subject might appear a violation of the principle of non-contradiction. This topic will not be further discussed here.

10. See J. Ladyman, 'Ontological, Epistemological, and Methodological Positions', in T. Kuipers (ed.), *General Philosophy of Science: Focal Issues* (Amsterdam: Elsevier, 2007), pp. 303–76, for a general introduction to the debate.

11. See, in particular, R. Giere, *Explaining Science: A Cognitive Approach* (Chicago, IL: University of Chicago Press, 1988), ch. 3; R. Giere, 'An Agent-Based Conception of Models and Scientific Representation', *Synthese*, 172 (2010), pp. 269–81; and M. S. Morgan and M. Morrison (eds), *Models as Mediators: Perspectives on Natural and Social Science* (Cambridge: Cambridge University Press, 1999), ch. 2.

12. This role has already been acknowledged in T. S. Kuhn, 'The Function of Measurement in Modern Physical Sciences', *Isis*, 52 (1961), pp. 161–93, and P. Suppes, 'Models of Data', in E. Nagel, P. Suppes and A. Tarski (eds), *Logic, Methodology, and Philosophy of Science* (Stanford, CA: Stanford University Press, 1962), pp. 252–61.

13. The introduction of models as mediators between the conceptualization of an entity and the entity itself allows reproduction of the opposition realism versus instrumentalist at the level of the theoretical entities whose existence is required in the construction of the model. For an extensive discussion of this issue in the recent epistemological debate see S. Psillos, *Scientific Realism: How Science Tracks the Truth* (London: Routledge, 1999), where scientific realism is defended, and B. C. van Fraassen, *Laws and Symmetry*

(Oxford: Oxford University Press, 1989), where a strong version of the opposite position, constructive empiricism, is developed.

14. JCGM 100, *Evaluation of Measurement Data*.

15. The Supplement 1 of the GUM, JCGM 101, *Evaluation of Measurement Data – Supplement 1 to the 'Guide to the Expression of Uncertainty in Measurement' – Propagation of Distributions using a Monte Carlo Method* (Joint Committee for Guides in Metrology, 2008), at http://www.bipm.org/en/publications/guides/gum.html [accessed 4 October 2013], acknowledges that this hypothesis can be generalized, and deals with measurement results in the form of probability distributions.

16. JCGM 200, *International Vocabulary of Metrology*.

17. See JCGM 100, *Evaluation of Measurement Data*, 2.2.4. According to the first of such 'other concepts', uncertainty would be 'a measure of error', a not-so-clear interpretation (e.g. is error considered here to be a non-quantitative concept, quantitatively expressed as/by means of uncertainty?). The subject will not further discussed here.

18. JCGM 200, *International Vocabulary of Metrology*, 2.3.

19. The concept was introduced as 'intrinsic uncertainty' in the GUM (and, as mentioned above, the first source of uncertainty listed by the GUM is indeed the 'incomplete definition of the measurand'), then re-termed 'definitional uncertainty' by the VIM3, JCGM 200, *International Vocabulary of Metrology*, 2.27, n. 3.

20. Ibid., 2.13.

21. ISO 5725–1, *Accuracy (Trueness and Precision) of Measurement Methods and Results – Part 1: General Principles and Definitions* (International Organization for Standardization, 1998).

22. As in the standard example: it is true that 'the snow is white' if and only if the snow is white, see M. David, 'The Correspondence Theory of Truth', in E. N. Zalta (ed.), *The Stanford Encyclopedia of Philosophy*, Fall 2009 Edition, at http://plato.stanford.edu/archives/fall2009/entries/truth-correspondence [accessed 4 October 2013], where 'the snow is white' and 'it is true that the snow is white' convey exactly the same information, and all usages of 'true' are of this kind, so that any reference to truth can be eliminated, D. Stoljar and N. Damnjanovic, 'The Deflationary Theory of Truth', in E. N. Zalta (ed.), *The Stanford Encyclopedia of Philosophy*, Winter 2010 Edition, at http://plato.stanford.edu/archives/win2010/entries/truth-deflationary [accessed 4 October 2013].

23. JCGM 200, *International Vocabulary of Metrology*, 3.1.

24. Ibid., 3.3.

25. Ibid., 4.1.

26. As the example suggests, the individual quantities q_{in} and q'_{out} are generally of different kinds, i.e. they are instances of different general quantities, Q and Q', where *in* is the object under measurement and *out* is usually the transducer itself, which changes its state as the result of its interaction with the object *in*.

27. Surprisingly, there seems to be some confusion even on this topic. For example, J. P. Bentley, *Principles of Measurement Systems* (Harlow: Pearson, 2005), p. 3, states: 'The input to the measurement system is the true value of the variable; the system output is the measured value of the variable. In an ideal measurement system, the measured value would be equal to the true value.' The assumption that the input of a measurement system is a quantity value, instead of a quantity, appears to be an obvious mistake.

28. For example, JCGM 104, *Evaluation of Measurement Data – An Introduction to the 'Guide to the Expression of Uncertainty in Measurement' and Related Documents* (Joint Committee for Guides in Metrology, 2009) states that 'in future editions of [the VIM]

it is intended to make a clear distinction between the use of the term "error" as a quantity and as a quantity value. The same statement applies to the term "indication". In the current document such a distinction is made. [On the other hand, the VIM3] does not distinguish explicitly between these uses'. The reference is plausibly to some inconsistencies in the VIM3, as when it is stated that 'indications, corrections and influence quantities can be input quantities in a measurement model', thus assuming indications as quantities, against the definition of 'indication' as 'quantity value provided by a measuring instrument or a measuring system'.

29. 'In carrying out measurements there is a tendency to reduce the immediate sensory observations, which of course can never be eliminated, to the safest and most exact among them, namely spatio-temporal coincidences (in particular, one tries to do without the subjective comparison of colors and light intensities). Any mensuration should ultimately ascertain, so one wishes, whether a mark on one scale (a movable pointer or such) coincides with a certain mark on another scale', H. Weyl, *Philosophy of Mathematics and Natural Science* (Princeton, NJ: Princeton University Press, 1949), p. 145.

30. See e.g. the quotation from Bentley, *Principles of Measurement Systems*, in n. 27, above.

31. Apparently this model does not comply with the customary condition that measurement implies comparison of quantities. In fact, in some cases measurement is based on the synchronous comparison of the object under measurement and a measurement standard, paradigmatically as performed by a two-pan balance. The simplified model presented here refers to the operatively much more frequent situation in which the object under measurement and the measurement standard are not required to be simultaneously present, under the hypothesis that the measuring instrument is stable enough to maintain in measurement the behaviour which was characterized in calibration. In this sense, the model is about measurement performed as asynchronous comparison of the object under measurement and the measurement standard.

32. JCGM 200, *International Vocabulary of Metrology*, 2.44.

33. Ibid., 2.15.

34. Ibid., 2.14.

35. The distinction in the VIM3 between trueness and accuracy, defined as 'closeness of agreement between a measured quantity value and a true quantity value of a measurand' (JCGM 200, *International Vocabulary of Metrology*, 2.13), is not completely clear, given that true quantity values are specific kinds of reference quantity values. In a different context (ISO, *Accuracy (Trueness and Precision) of Measurement Methods and Results*), building on measurement precision and trueness, measurement accuracy is thought of as an overall indicator, which summarizes the information conveyed by both precision and trueness. How such information can be synthesized is outside the scope of this chapter.

5 Beck, 'Handling Uncertainty in Environmental Models at the Science–Policy–Society Interfaces

1. A. Saltelli, K. Chan and E. M. Scott (eds), *Sensitivity Analysis* (Chichester: Wiley, 2000).

2. E.g. R. V. O'Neill, 'Error Analysis of Ecological Models', in *Radionuclides in Ecosystems*, Conf 710501 (Springfield, VA: National Technical Information Service, 1973), pp. 898–908; M. B. Beck and G. van Straten (eds), *Uncertainty and Forecasting of Water Quality* (Berlin: Springer, 1983); L. C. Brown and T. O. Barnwell, 'The Enhanced Stream Water Quality Models QUAL2E and QUAL2E-UNCAS: Documentation and User

Model', *Report EPA/600/3-87/007* (Athens, GA: United States Environmental Protection Agency, 1987); M. B. Beck, 'Water Quality Modeling: A Review of the Analysis of Uncertainty', *Water Resources Research*, 23:8 (1987), pp. 1393–442; M. G. Morgan and M. Henrion, *Uncertainty: A Guide to Dealing with Uncertainty in Quantitative Risk and Policy Analysis* (Cambridge: Cambridge University Press, 1990).

3. See, for example, J. P. van der Sluijs, 'Uncertainty and Precaution in Environmental Management: Insights from the UPEM Conference', *Environmental Modelling & Software*, 22 (2007), pp. 590–8, and J. A. Curry and P. J. Webster, 'Climate Science and the Uncertainty Monster', *Bulletin American Meteorological Society*, December (2011), pp. 1667–82.

4. NRC, *Models in Environmental Regulatory Decision Making* (Washington, DC: National Research Council and National Academy Press, 2007).

5. Ibid.; W. Wagner, E. Fisher and P. Pascual, 'Misunderstanding Models in Environmental and Public Health Regulation', *New York University Law Journal*, 18 (2010), pp. 293–356; E. Fisher, P. Pascual and W. Wagner, 'Understanding Environmental Models in their Legal and Regulatory Context', *Journal of European Environmental Law*, 22:2 (2010), pp. 251–83.

6. NRC, *Models in Environmental Regulatory Decision Making*, p. 125.

7. O. H. Pilkey and L. Pilkey-Jarvis, *Useless Arithmetic: Why Environmental Scientists Can't Predict the Future* (New York: Columbia University Press, 2007), and the rebuttal, in effect: M. B. Beck, 'How Best to Look Forward', *Science*, 316 (2007), pp. 202–3.

8. Uncertainty and error in collecting the data (metrology) are the subject of the chapter by Mari and Giordani, this volume. Measurements, measuring instruments and models are the subject of Box 1, pp. 103–5, this chapter. Its discussion of model calibration, verification, validation and (not least) whether we should place our trust in any data 'speaking for themselves', is important to reading the present chapter in the light of Mari and Giordani's chapter.

9. As in M. Kandlikar, J. Risbey and S. Dessai, 'Representing and Communicating Deep Uncertainty in Climate-Change Assessments', *Comptes Rendus Geoscience*, 337 (2005), pp. 443–55.

10. See also Wagner, Fisher and Pascual, 'Misunderstanding Models in Environmental and Public Health Regulation'; Fisher, Pascual and Wagner, 'Understanding Environmental Models in their Legal and Regulatory Context'.

11. S. O. Funtowicz and J. R. Ravetz, *Uncertainty and Quality in Science for Policy* (Dordrecht: Kluwer, 1990).

12. S. Jasanoff, *The Fifth Branch: Science Advisors as Policymakers* (Cambridge, MA: Harvard University Press, 1990).

13. M. Thompson, R. Ellis and A. Wildavsky, *Cultural Theory* (Boulder, CO: West View, 1990).

14. M. S. Morgan, *The World in the Model: How Economists Work and Think* (Cambridge: Cambridge University Press, 2012).

15. Wagner, Fisher and Pascual, 'Misunderstanding Models in Environmental and Public Health Regulation'.

16. Ibid.

17. M. B. Beck, 'Uncertainty, System Identification and the Prediction of Water Quality', in Beck and van Straten (eds), *Uncertainty and Forecasting of Water Quality*, pp. 3–68; M. B. Beck (ed.), *Environmental Foresight and Models: A Manifesto* (Oxford: Elsevier, 2002); M. B. Beck and E. Halfon, 'Uncertainty, Identifiability and the Propagation of Prediction Errors: A Case Study of Lake Ontario', *Journal of Forecasting*, 10:1–2 (1991), pp. 135–61; M. B. Beck, A. J. Jakeman and M. J. McAleer, 'Construction and Evaluation

of Models of Environmental Systems', in A. J. Jakeman, M. B. Beck and M. J. McAleer (eds), *Modelling Change in Environmental Systems* (Chichester: Wiley, 1993), pp. 3–35; M. B. Beck, et al., *Grand Challenges of the Future for Environmental Modeling, White Paper* (Arlington, VA: National Science Foundation, 2009).

18. Beck, et al., *Grand Challenges of the Future for Environmental Modeling*; M. B. Beck, Z. Lin and J. D. Stigter, 'Model Structure Identification and the Growth of Knowledge', in L. Wang and H. Garnier (eds), *System Identification, Environmental Modelling, and Control System Design* (London: Springer, 2011), pp. 69–96.

19. NRC, *Models in Environmental Regulatory Decision Making*.

20. As elaborated in Curry and Webster 'Climate Science and the Uncertainty Monster'.

21. Beck, Lin and Stigter, 'Model Structure Identification and the Growth of Knowledge'.

22. According to Beck, et al., *Grand Challenges of the Future for Environmental Modeling*.

23. Beck, 'Uncertainty, System Identification and the Prediction of Water Quality'; Beck, 'Water Quality Modeling'.

24. Where the authenticity and veracity of these references derive their legitimacy from a chain of logic leading all the way back to the ultimate referencing devices, such as an atomic clock or a platinum-iridium rod of length one meter.

25. There are wheels within wheels, the mention of which is best kept from the mainstream discussion, in a footnote, in a text box, itself to one side of the main course of the discussion of this chapter. These inner wheels have to do with the terms 'verification' and 'validation'. For Mari and Giordani, the verification of an instrument comes with the 'provision of objective evidence that a given item fulfils specified requirements'. For the present chapter, verification means that the model M has been shown to match one set of field data, say $[u, y]_1$. Validation, according to its common usage (now questioned, as this chapter shows), is the act of demonstrating that model M matches not only $[u, y]_1$, but also a second set of independently gathered-in data $[u, y]_2$, subject to *no* change in the values assigned to the model's parameters (α) during calibration of M to $[u, y]_1$, if such prior calibration was exercised with respect to $[u, y]_1$, see M. B. Beck, 'Model Evaluation and Performance', in A. H. El-Shaarawi and W. W. Piegorsch (eds), *Encyclopedia of Environmetrics* (Chichester: Wiley, 2002), vol. 3, pp. 1275–9. Ideally, the range of empirical patterns of behaviour captured in $[u, y]_2$ might be different from those captured in $[u, y]_1$.

26. Modern-day respirometers, fully intended as measuring devices, are themselves literally instrumented microcosms of complex microbiological ecosystems, see P. A. Vanrolleghem, et al., 'Estimating (Combinations of) Activated Sludge Model No. 1 Parameters and Components by Respirometry', *Water Science and Technology*, 39:1 (1999), pp. 195–214.

27. M. B. Beck, 'Model Structure Identification from Experimental Data', in E. Halfon (ed.), *Theoretical Systems Ecology. Advances and Case Studies* (New York: Academic, 1979), pp. 259–89.

28. Beck (ed.), *Environmental Foresight and Models*; O. O. Osidele and M. B. Beck, 'An Inverse Approach to the Analysis of Uncertainty in Models of Environmental Systems', *Integrated Assessment*, 4:4 (2003), pp. 265–83.

29. See originally, G. M. Hornberger and R. C. Spear, 'Eutrophication in Peel Inlet, I, Problem-defining Behaviour and a Mathematical Model for the Phosphorus Scenario', *Water Research*, 14 (1980), pp. 29–42; and R. C. Spear and G. M. Hornberger, 'Eutrophication in Peel Inlet, II, Identification of Critical Uncertainties via Generalized Sensitivity Analysis', *Water Research*, 14 (1980), pp. 43–9.

30. Beck (ed.), *Environmental Foresight and Models*; M. B. Beck, 'Environmental Foresight and Structural Change', *Environmental Modelling & Software*, 20:6 (2005), pp. 651–70; Z. Lin and M. B. Beck, 'Accounting for Structural Error and Uncertainty in a Model: An Approach Based on Model Parameters as Stochastic Processes', *Environmental Modelling & Software*, 27–8 (2012), pp. 97–111.

31. Beck (ed.), *Environmental Foresight and Models*; Beck, 'Environmental Foresight and Structural Change'; Beck, et al., *Grand Challenges of the Future for Environmental Modeling*; Box 1; and Mari and Giordani, this volume.

32. The reasons for this rewording have to do with the culture of the environmental sciences, the change being chronicled over the years in, amongst others, L. F. Konikow and J. D. Bredehoeft, 'Groundwater Models Cannot be Validated', *Advances in Water Resources*, 15:1 (1992), pp. 75–83; N. Oreskes, K. Shrader-Frechette and K. Belitz, 'Verification, Validation, and Confirmation of Numerical Models in the Earth Sciences', *Science*, 263 (1994), pp. 641–6; M. B. Beck, J. R. Ravetz, L. A. Mulkey and T. O. Barnwell, 'On the Problem of Model Validation for Predictive Exposure Assessments', *Stochastic Hydrology and Hydraulics*, 11:3 (1997), pp. 229–54; N. Oreskes, 'Evaluation (Not Validation) of Quantitative Models', *Environmental Health Perspectives*, 106: Supplement 6 (1998), pp. 1453–60; M. B. Beck and J. Chen, 'Assuring the Quality of Models Designed for Predictive Tasks', in Saltelli, Chan and Scott (eds), *Sensitivity Analysis*, pp. 401–20; and Beck, 'Model Evaluation and Performance'.

33. Beck, Lin and Stigter, 'Model Structure Identification and the Growth of Knowledge'.

34. M. P. Krayer von Krauss, et al., 'Uncertainty and Precaution in Environmental Management', *Water Science and Technology*, 52:6 (2005), pp. 1–9.

35. M. B. Beck, 'Water Quality', *Journal of the Royal Statistical Society, Series A (General)*, 147:2 (1984), pp. 293–305.

36. And in any case, other EU countries were able to exploit their geographical setting to be at the heart of Europe's transport system or to grow citrus fruits, so why should the UK not benefit from its natural advantages of location?

37. Wagner, Fisher and Pascual, 'Misunderstanding Models in Environmental and Public Health Regulation'; Fisher, Pascual and Wagner, 'Understanding Environmental Models in their Legal and Regulatory Context'. Another was to bring scholarship from the social and political sciences, including psychology, and the perspective of the media, to bear on the same. In particular, participants in TAUC from both sides of the Atlantic joined forces to examine the empirical evidence (in US public-policy administration) of updating and adapting policies, rules and regulations in the light of ever-evolving knowledge bases, see L. E. McCray, K. A. Oye and A. C. Petersen, 'Planned Adaptation in Risk Regulation: An Initial Survey of US Environmental, Health, and Safety Regulation', *Technological Forecasting & Social Change*, 77 (2010), pp. 951–9.

38. Wagner, Fisher and Pascual, 'Misunderstanding Models in Environmental and Public Health Regulation'.

39. Fisher, Pascual and Wagner, 'Understanding Environmental Models in their Legal and Regulatory Context'.

40. As in T. N. Palmer and P. J. Hardaker, 'Handling Uncertainty in Science', *Philosophical Transactions Royal Society A*, 369 (2011), pp. 4681–4.

41. Beck, et al., *Grand Challenges of the Future for Environmental Modeling*.

42. P. R. Moorcroft, 'How Close are We to a Predictive Science of the Biosphere?', *TRENDS in Ecology and Evolution*, 21:7 (2006), pp. 400–7.

43. A. Patt, 'Assessing Model-Based and Conflict-Based Uncertainty', *Global Environmental Change*, 17 (2007), pp. 37–46.

44. Funtowicz and Ravetz, *Uncertainty and Quality in Science for Policy*.

45. Kuhn, *The Structure of Scientific Revolutions* (1962).

46. Funtowicz and Ravetz, *Uncertainty and Quality in Science for Policy*.

47. M. B. Beck, et al., 'Developing a Concept of Adaptive Community Learning: Case Study of a Rapidly Urbanizing Watershed', *Integrated Assessment*, 3:4 (2002), pp. 299–307; M. P. Hare, et al., in C. Giupponi, A. J. Jakeman, D. Karssenberg and M. P. Hare (eds), *Sustainable Management of Water Resources: An Integrated Approach* (Northampton, MA: Edward Elgar, 2006), pp. 177–231; and M. B. Beck, et al., 'On Governance for Re-engineering City Infrastructure', *Proceedings of the Institution of Civil Engineers, Engineering Sustainability*, 164:ES2 (2011), pp. 129–42.

48. A. C. Petersen, et al., 'Post-Normal Science in Practice at the Netherlands Environmental Assessment Agency', *Science, Technology, & Human Values*, 36:3 (2011), pp. 362–88.

49. S. Ney, *Resolving Messy Policy Problems: Handling Conflict in Environmental, Transport, Health and Ageing Policy* (London: Earthscan, 2009).

50. R. A. Dahl, *Democracy and Its Critics* (New Haven, CT: Yale University Press, 1989).

51. The original statement in the 2007 IPCC Assessment regarding disappearance of the Himalayan glaciers by 2035, was erroneous, having been drawn from a World Wide Fund for Nature report, which had cited a news story, which in turn had drawn upon an unpublished study that estimated *no* date for any such 'disappearance', PBL, *Assessing an IPCC Assessment: An Analysis of Statements on Projected Regional Impacts in the 2007 Report* (The Hague and Bilthoven: Netherlands Environmental Assessment Agency (PBL), 2010), Annex B, pp. 86–8.

52. Thompson, Ellis and Wildavsky, *Cultural Theory*.

53. Kuhn, *The Structure of Scientific Revolutions*.

54. Jasanoff, *The Fifth Branch*.

55. O. Klepper, H. Scholten and J. P. G. van de Kamer, 'Prediction Uncertainty in an Ecological Model of the Oosterschelde Estuary', *Journal of Forecasting*, 10:1–2 (1991), pp. 191–209.

56. Ibid.

57. See, for example, D. von Winterfeld and W. Edwards, *Decision Analysis and Behavioural Research* (Cambridge: Cambridge University Press, 1986).

58. For example, M. P. Krayer von Krauss and P. H. M. Janssen, 'Using the W&H Integrated Uncertainty Analysis Framework with Non-initiated Experts', *Water Science and Technology*, 52:6 (2005), pp. 145–52; and as adumbrated in Kandlikar, Risbey and Dessai, 'Representing and Communicating Deep Uncertainty in Climate-change Assessments', as well as further elaborated in Beck, et al., *Grand Challenges of the Future for Environmental Modeling*.

59. Funtowicz and Ravetz, *Uncertainty and Quality in Science for Policy*.

60. According to M. B. A. van Asselt and J. Rotmans, 'Uncertainty in Integrated Assessment Modelling: From Positivism to Pluralism', *Climatic Change*, 54 (2002), pp. 75–105.

61. Curry and Webster, 'Climate Science and the Uncertainty Monster'.

62. M. Thompson, 'Decision Making under Contradictory Certainties: How to Save the Himalayas When You Can't Find Out What's Wrong with Them', *Applied Systems Analysis*, 12 (1985), pp. 3–34.

63. Funtowicz and Ravetz, *Uncertainty and Quality in Science for Policy*.

64. Thompson, Ellis and Wildavsky, *Cultural Theory*.

65. M. Nowacki, S. Luan and M. Verweij, 'I Disagree, Therefore I Am; Moral Discord Among Children', *International Workshop on The Human Brain and the Social Bond: Exploring the Notion of Constrained Relativism* (Altenberg and Laxenburg: Konrad Lorenz Institute and International Institute for Applied Systems Analysis, 2010).

66. Ney, *Resolving Messy Policy Problems*.

67. M. B. A. van Asselt and J. Rotmans, 'Uncertainty in Perspective', *Global Environmental Change*, 6:2 (1996), pp. 121–57.

68. Funtowicz and Ravetz, *Uncertainty and Quality in Science for Policy*.

69. C. Coglianese, 'Is Consensus an Appropriate Basis for Regulatory Policy?', in E. Orts and K. Deketelaere (eds), *Environmental Contracts: Comparative Approaches to Regulatory Innovation in the United States and Europe* (Dordrecht: Kluwer, 2001), pp. 93–113; see also M. B. Beck, *Cities as Forces for Good in the Environment: Sustainability in the Water Sector* (Athens, GA: Warnell School of Forestry and Natural Resources, University of Georgia, 2011).

70. Funtowicz and Ravetz, *Uncertainty and Quality in Science for Policy*.

71. S. O. Funtowicz and J. R. Ravetz, 'A New Scientific Methodology for Global Environmental Issues', in R. Costanza (ed.), *Ecological Economics: The Science and Management of Sustainability* (New York: Columbia University Press, 1991), pp. 137–52.

72. For example Petersen, et al., 'Post-Normal Science in Practice at the Netherlands Environmental Assessment Agency'.

73. McCray, Oye and Petersen, 'Planned Adaptation in Risk Regulation'.

74. According to K. R. Foster, P. Vecchia and M. H. Repacholi, 'Science and the Precautionary Principle', *Science*, 288:5468 (2000), pp. 979–81.

75. A. James (ed.), *Mathematical Models in Water Pollution Control* (Chichester: Wiley, 1978); G. T. Orlob (ed.), *Mathematical Modeling of Water Quality: Streams, Lakes, and Reservoirs* (Chichester: Wiley, 1983).

76. Thompson, Ellis and Wildavsky, *Cultural Theory*.

77. M. A. Janssen and S. R. Carpenter, 'Managing the Resilience of Lakes: A Multi-Agent Modeling Approach', *Conservation Ecology*, 3:2 (1999). See also Box 6 (pp. 82–4) in Beck, et al., *Grand Challenges of the Future for Environmental Modeling*, which charts some preliminary responses to challenge no. 10 for the future of environmental modelling, i.e. in the use of models for management and decision support.

78. James, *Mathematical Models in Water Pollution Control*; Orlob, *Mathematical Modeling of Water Quality*.

79. L. J. Mulkey, 'A De Novo Earth System Model', *Internal Report* (Athens, GA: United States Environmental Protection Agency, 1991).

80. Van der Sluijs, 'Uncertainty and Precaution in Environmental Management'.

81. Thompson, Ellis and Wildavsky, *Cultural Theory*.

82. See also A. C. Petersen, *Simulating Nature: A Philosophical Study of Computer-Simulation Uncertainties and Their Role in Climate Science and Policy Advice* (Apeldoorn: Het Spinhuis, 2006); C. Mooney, *Storm World – Hurricanes, Politics, and the Battle Over Global Warming* (Orlando, FL: Harcourt, 2007); T. O. McGarity and W. E. Wagner, *Bending Science. How Special Interests Corrupt Public Health Research* (Cambridge, MA: Harvard University Press, 2008); N. Oreskes and E. M. Conway, *Merchants of Doubt* (London: Bloomsbury Press, 2010).

83. Wagner, Fisher and Pascual, 'Misunderstanding Models in Environmental and Public Health Regulation'.

84. Mulkey, 'A De Novo Earth System Model'.

85. Morgan, *The World in the Model*.
86. The robust opposition of the advocates of the two – large and small *M* – is recorded in the proceedings of a 1974 International Federation of Information Processing (IFIP) 'Working Conference on Computer Simulation of Water Resources Systems', G. C. Vansteenkiste (ed.), *Computer Simulation of Water Resources Systems* (Amsterdam: North-Holland, 1975). Standing somewhat above the fray, and using 'uncertainty' to do so, one might recast this opposition as a matter of both being distant from perfection: with a large model, one might be able to predict a correct future, but not know this, in the midst of all the uncertainty; with a small model, one might predict a quite incorrect future and, worse still, be most confident in so doing, see Beck, 'Uncertainty, System Identification and the Prediction of Water Quality'. Today, the two poles of the duality can be seen to have mutually reinforcing and substantial benefits, see Z. Lin and M. B. Beck, 'Understanding Complex Environmental Systems: A Dual Thrust', *Environmetrics*, 18:1 (2007), pp. 11–26; Beck, et al., *Grand Challenges of the Future for Environmental Modeling*.
87. Funtowicz and Ravetz, *Uncertainty and Quality in Science for Policy*.
88. Wagner, Fisher and Pascual, 'Misunderstanding Models in Environmental and Public Health Regulation'.
89. Beck (ed.), *Environmental Foresight and Models*.
90. B. J. Mason (ed.), *The Surface Waters Acidification Programme* (Cambridge: Cambridge University Press, 1990).
91. Pilkey and Pilkey-Jarvis, *Useless Arithmetic*.
92. M. Crichton, *State of Fear* (New York: Avon, 2004), p. 628.
93. J. P. van der Sluijs, 'A Way Out of the Credibility Crisis for Models Used in Integrated Environmental Assessment', *Futures*, 34 (2002), pp. 133–46.
94. Petersen, et al., 'Post-Normal Science in Practice at the Netherlands Environmental Assessment Agency'.
95. Ibid.
96. Beck, 'How Best To Look Forward'.
97. S. Rayner, as reported in the Record of a Workshop on 'A New Look at the Interaction of Scientific Models and Policymaking', Policy Foresight Programme, The James Martin 21st Century School, University of Oxford, UK, 2008, at http://www.martininstitute. ox.ac.uk/jmi/networks/Policy+Foresight+Programme.htm [accessed 8 July 2009].
98. S. Schaffer, 'Comets and the World's End', in L. Howe and A. Wain (eds), *Predicting the Future* (Cambridge: Cambridge University Press, 1993), pp. 52–76.
99. K. C. Green and J. S. Armstrong, 'Global Warming: Forecasts by Scientists versus Scientific Forecasts', *Energy and Environment*, 18:7–8 (2007), pp. 997–1021.
100. G. J. Jenkins, J. M. Murphy, D. M. H. Sexton, J. A. Lowe, P. Jones and C. G. Kilsby, *UK Climate Projections: Briefing Report* (Exeter, UK: Meteorological Office Hadley Centre, 2009), version 2, December 2010, at http://ukclimateprojections.defra.gov.uk [accessed 16 March 2013].
101. EEA, *Late Lessons from Early Warnings: The Precautionary Principle 1896–2000*, Environmental Issue Report 22 (Copenhagen: European Environment Agency, 2001); EEA, *Late Lessons from Early Warnings: Science, Precaution, Innovation*, EEA Report 1/2013 (Copenhagen: European Environment Agency, 2013).
102. M. B. Beck, 'The Manifesto', in Beck (ed.), *Environmental Foresight and Models*, pp. 61–93.
103. I emphasize the '*I*' here, because, first, I wish to make this statement strictly personal, thereby deliberately exposing it as such for public scrutiny, because it may be far from the

general perception in these matters. Second, I wish to draw attention to another change over the past fifty years: from the detached, supposedly objective, third-person singular and passive, to the first-person singular and active authorship in scientific writing.

104. See the report on the 13 February 2008, Seminar 'A New Look at the Interaction of Scientific Models with Policy Making', held within the Policy Foresight Programme of the James Martin Institute at the University of Oxford, UK (www.martininstitute.ox.ac. uk, [accessed 8 July 2009]). The comments of Rayner reside in this report. Leonard A. Smith, co-author, with A. C. Petersen, of Chapter 6 in this volume, also participated in the seminar. Thus in this latter chapter we find that climate scientists should henceforth be 'overhumble' in order to regain the public's trust; they should stoutly resist any temptation to hide the 'relevant dominant uncertainties' in their works.

105. Beck, 'Water Quality Modeling'; Beck and Halfon, 'Uncertainty, Identifiability and the Propagation of Prediction Errors'.

106. Beck (ed.), *Environmental Foresight and Models*; Beck 'Environmental Foresight and Structural Change'; Beck, et al., *Grand Challenges of the Future for Environmental Modeling*.

107. Beck, 'Model Evaluation and Performance'.

108. Beck, Lin and Stigter, 'Model Structure Identification and the Growth of Knowledge'.

109. P. Reichert and J. Mieleitner, 'Analyzing Input and Structural Uncertainty of Nonlinear Dynamic Models with Stochastic, Time-dependent Parameters', *Water Resources Research*, 45 (2009), W10402.

110. L. Tomassini, et al., 'A Smoothing Algorithm for Estimating Stochastic, Continuous Time Model Parameters and its Application to a Simple Climate Model', *Journal of the Royal Statistical Society, Series C*, 58:5 (2009), pp. 679–704.

111. Beck and Halfon, 'Uncertainty, Identifiability and the Propagation of Prediction Errors'.

112. K. J. Beven and J. Freer, 'Equifinality, Data Assimilation, and Uncertainty Estimation in Mechanistic Modelling of Complex Environmental Systems', *Journal of Hydrology*, 249 (2001), pp. 11–29.

113. See Lin and Beck, 'Accounting for Structural Error and Uncertainty in a Model'.

114. Morgan, *The World in the Model*.

115. M. B. Beck, J. Chen and O. O. Osidele, 'Random Search and the Reachability of Target Futures', in Beck (ed.), *Environmental Foresight and Models*, pp. 207–26; Osidele and Beck, 'An Inverse Approach to the Analysis of Uncertainty in Models of Environmental Systems'.

116. Beck, 'The Manifesto'.

117. Beck, *Cities as Forces for Good in the Environment*; M. B. Beck, R. Villarroel Walker and M. Thompson, 'Smarter Urban Metabolism: Earth Systems Re-Engineering', *Proceedings of the Institution of Civil Engineers, Engineering Sustainability*, 163(ES5) (2013), pp. 229–41.

118. Osidele and Beck, 'An Inverse Approach to the Analysis of Uncertainty in Models of Environmental Systems'.

119. K. J. Keesman, 'Parametric Change as the Agent of Control', in Beck (ed.), *Environmental Foresight and Models*, pp. 415–24; A. V. Kryazhimskii and M. B. Beck, 'Identifying the Inclination of a System Towards a Terminal State from Current Observations', in Beck (ed.), *Environmental Foresight and Models*, pp. 425–51.

120. Lin and Beck, 'Accounting for Structural Error and Uncertainty in a Model'.

121. Beck, Lin and Stigter, 'Model Structure Identification and the Growth of Knowledge'.

122. See Wagner, Fisher and Pascual, 'Misunderstanding Models in Environmental and Public Health Regulation'; Fisher, Pascual and Wagner, 'Understanding Environmental Models in their Legal and Regulatory Context'.

123. As discussed at length elsewhere; Beck, et al., 'On the Problem of Model Validation for Predictive Exposure Assessments'; Beck and Chen, 'Assuring the Quality of Models Designed for Predictive Tasks'; Beck, *Environmental Foresight and Models*; Beck, 'Model Evaluation and Performance'; NRC, *Models in Environmental Regulatory Decision Making*.

124. Beck and Chen, 'Assuring the Quality of Models Designed for Predictive Tasks'.

125. Beck, 'Model Evaluation and Performance'; Beck, et al., 'On the Problem of Model Validation for Predictive Exposure Assessments'; Funtowicz and Ravetz, *Uncertainty and Quality in Science for Policy*.

126. Wagner, Fisher and Pascual, 'Misunderstanding Models in Environmental and Public Health Regulation'; Fisher, Pascual and Wagner, 'Understanding Environmental Models in their Legal and Regulatory Context'.

127. Wagner, Fisher and Pascual, 'Misunderstanding Models in Environmental and Public Health Regulation'.

128. Ibid.

129. M. Thompson, *Organising and Disorganising: A Dynamic and Non-linear Theory of Institutional Emergence and Its Implications* (Axminster: Triarchy, 2008); Ney, *Resolving Messy Policy Problems*.

130. C. S. Holling (ed.), *Adaptive Environmental Assessment and Management* (Chichester: Wiley, 1978).

131. M. Thompson, 'Man and Nature as a Single but Complex System', in P. Timmerman (ed.), *Encyclopedia of Global Environmental Change*, vol. 5 (Chichester: Wiley, 2002), pp. 384–93.

132. Wagner, Fisher and Pascual, 'Misunderstanding Models in Environmental and Public Health Regulation'.

133. Beck, et al., 'Developing a Concept of Adaptive Community Learning'.

134. Osidele and Beck, 'An Inverse Approach to the Analysis of Uncertainty in Models of Environmental Systems'.

135. Thompson, *Organising and Disorganising*; Ney, *Resolving Messy Policy Problems*.

136. In Beck, et al., 'Model Structure Identification and the Growth of Knowledge', in L. Wang and H. Garnier (eds), *System Identification, Environmental Modelling, and Control System Design*; and Beck *Cities as Forces for Good in the Environment*.

137. Beck, *Cities as Forces for Good in the Environment*.

138. Wagner, Fisher and Pascual, 'Misunderstanding Models in Environmental and Public Health Regulation'; Fisher, Pascual and Wagner, 'Understanding Environmental Models in their Legal and Regulatory Context'.

139. Wagner, Fisher and Pascual, 'Misunderstanding Models in Environmental and Public Health Regulation'.

140. Ibid.

141. Thompson, *Organising and Disorganising*; Ney, *Resolving Messy Policy Problems*.

142. M. H. Shapiro, 'Introduction: Judicial Selection and the Design of Clumsy Institutions', *Southern California Law Review*, 61 (1988), pp. 1555–69.

143. Beck, et al., 'Developing a Concept of Adaptive Community Learning'; Beck, *Cities as Forces for Good in the Environment*.

6 Smith and Petersen, 'Variations on Reliability: Connecting Climate Predictions to Climate Policy'

1. We would like to thank Mike Hulme for his suggestion in June 2010 to add the notion of reliability$_3$ to the notions of reliability$_1$ and reliability$_2$, that were first introduced in A. C. Petersen, *Simulating Nature: A Philosophical Study of Computer-Simulation Uncertainties and Their Role in Climate Science and Policy Advice*, 2nd edn (Boca Raton, FL: CRC Press, [2006] 2012), and Brian Hoskins, Wendy Parker, Dave Stainforth and Nick Stern for continuing conversations on the matters of this chapter, and beyond.

2. The distinction between 'prediction' and 'projection' is arguably artificial if not simply false. As discussed below all probability forecasts, indeed all probability statements, are conditional on some information.

3. Our focus in this chapter is on 'today's science' and questions that might, or might not, be answered via a probability distribution. A. M. Weinberg, 'Science and Trans-Science', *Minerva*, 10 (1972), pp. 209–22, used the term 'trans-science' to denote the situation of questions being asked of science that science cannot answer. S. O. Funtowicz and J. R. Ravetz, 'Science for the Post-Normal Age', *Futures*, 25 (1993), pp. 739–55, refer to 'post-normal science' as the appropriate problem-solving strategy for this situation given that one still wants to make responsible use of the scientific knowledge available, see also Petersen, et al., 'Post-Normal Science in Practice at the Netherlands Environmental Assessment Agency' and A. Millner, R. Calel, D. Stainforth and G. MacKerron, 'Do Probabilistic Expert Elicitations Capture Scientists' Uncertainty about Climate Change?', *Climatic Change*, 116:2 (2013), pp. 427–36.

4. Statisticians have long recognized that confidence intervals on model-based forecasts are exceeded more often than theory suggests they should be. Reproducibility of a specific calculation, robustness of the statistics of a specific model, and belief in the fidelity of that model statistic (that it reflects reality) are three distinct things.

5. L. A. Smith and N. Stern, 'Scientific Uncertainty and Policy Making', *Philosophical Transactions of the Royal Society A*, 369 (2011), pp. 1–24.

6. RDU will be used as both singular and plural. It may be thought of as a set of uncertainties where one dominates, as a many-headed Hydra where, as the science advances and removes one head, another immediately replaces it.

7. See e.g. D. A. Randall and B. A. Wielicki, 'Measurements, Models, and Hypotheses in the Atmospheric Sciences', *Bulletin of the American Meteorological Society*, 78 (1997), pp. 399–406; J. P. van der Sluijs, *Anchoring amid Uncertainty: On the Management of Uncertainties in Risk Assessment of Anthropogenic Climate Change* (Utrecht: Utrecht University, 1997); R. Moss and S. Schneider, 'Uncertainties in the IPCC TAR: Recommendations to Lead Authors for More Consistent Assessment and Reporting', in R. Pachauri, T. Taniguchi and K. Tanaka (eds), *Guidance Papers on the Cross Cutting Issues of the Third Assessment Report of the IPCC* (Geneva, Switzerland: Intergovernmental Panel on Climate Change, 2000), available at http://climateknowledge.org/figures/Rood_Climate_Change_AOSS480_Documents/Moss_Schneider_Consistent_Reporting_Uncertainty_IPCC_2000.pdf [accessed 16 October 2013]; A. C. Petersen, 'Philosophy of Climate Science', *Bulletin of the American Meteorological Society*, 81 (2000), pp. 265–71; Petersen, *Simulating Nature*; IPCC, *Guidance Notes for Lead Authors of the IPCC Fourth Assessment Report on Addressing Uncertainties* (Geneva, Switzerland: Intergovernmental Panel on Climate Change, 2005), available at http://www.ipcc.ch/pdf/assessment-report/ar4/wg1/ar4-uncertaintyguidancenote.pdf [accessed

16 October 2013]; J. S. Risbey and M. Kandlikar, 'Expressions of Likelihood and Confidence in the IPCC Uncertainty Assessment Process', *Climatic Change*, 85 (2007), pp. 19–31; R. Swart, L. Bernstein, M. Ha-Duong and A. C. Petersen, 'Agreeing to Disagree: Uncertainty Management in Assessing Climate Change, Impacts and Responses by the IPCC', *Climatic Change*, 92 (2009), pp. 1–29; M. Hulme and M. Mahony, 'Climate Change: What Do We Know about the IPCC?', *Progress in Physical Geography*, 34 (2010), pp. 705–18; M. D. Mastrandrea, et al., *Guidance Note for Lead Authors of the IPCC Fifth Assessment Report on Consistent Treatment of Uncertainties* (Geneva, Switzerland: Intergovernmental Panel on Climate Change, 2010), available at http://www.ipcc. ch/pdf/supporting-material/uncertainty-guidance-note.pdf [accessed 16 October 2013].

8.   Ibid.

9.   Moss and Schneider, 'Uncertainties in the IPCC TAR'.

10. IAC, *Climate Change Assessments: Review of the Processes and Procedures of the IPCC* (Amsterdam: InterAcademy Council, 2010), available at http://reviewipcc.interacademycouncil.net [accessed 16 October 2013].

11. Petersen, *Simulating Nature*; IAC, *Climate Change Assessments*; B. J. Strengers, et al., *Opening up Scientific Assessments for Policy: The Importance of Transparency in Expert Judgements*, PBL Working Paper 14 (The Hague: PBL Netherlands Environmental Assessment Agency, 2013).

12. How might a consultant who communicates the RDU openly maintain the attention and interest of decision-makers? How is one to avoid the false claim that fair and needed scientific uncertainty (exposed by discussion of the RDU) suggests that the science is too uncertain to justify action at this point in time?

13. I. J. Good, *Good Thinking: The Foundations of Probability and Its Applications* (Mineola, NY: Dover Publications, [1983] 2009) noted that there are 46,656 varieties of Bayesian; UKCP09 may well have introduced at least one more variety.

14. See Kandlikar, Risbey and Dessai, 'Representing and Communicating Deep Uncertainty in Climate Change Assessments'; called 'ambiguity' in Smith and Stern, 'Scientific Uncertainty and Policy Making' and 'Knightian uncertainty' in economics.

15. See Petersen, *Simulating Nature*.

16. D. A. Stainforth, et al., 'Issues in the Interpretation of Climate Model Ensembles to Inform Decisions', *Philosophical Transactions of the Royal Society A Mathematical Physical and Engineering Sciences*, 365:1857 (2007), pp. 2163–77.

17. Petersen, *Simulating Nature*.

18. The finding that most of the observed warming over the last fifty years is attributable to the anthropogenic emissions of greenhouse gases was qualified in 2001 as 'likely' (defined as a 'judgmental estimate' of a 66–90 per cent chance of being correct) and in 2007 as 'very likely' (a > 90 per cent chance). How objective and subjective probabilities were combined in these uncertainty qualifications is analysed in detail by Petersen, *Simulating Nature*.

19. Good, *Good Thinking*.

20. L. A. Smith, 'What Might We Learn from Climate Forecasts?', *Proceedings of the National Academy of Sciences of the United States of America* 99, suppl. 1 (2002), pp. 2487–92; L. A. Smith, 'Predictability Past Predictability Present', in T. Palmer and R. Hagedorn (eds), *Predictability of Weather and Climate* (Cambridge, UK: Cambridge University Press, 2006).

21. Indeed, this applies to all simulation models as well as 'modes of thought' (see Whitehead's 'fallacy of misplaced concreteness').

22. Smith, 'What Might We Learn from Climate Forecasts?'
23. W. S. Parker, 'Confirmation and Adequacy-For-Purpose in Climate Modeling', *Proceedings of the Aristotelian Society* 83, suppl. (2009), pp. 233–49.
24. See Smith and Stern, 'Scientific Uncertainty and Policy Making' and L. A. Smith, 'Insight and Action: Probabilistic Prediction in Practice', *International Journal of Forecasting* (2013, forthcoming).
25. See Petersen, *Simulating Nature*.
26. And that of the scientists.
27. There are, of course, climate scientists who model and climate modellers who are scientists, the distinction here is meant to reflect those whose primary direct focus of interest is the Earth System from those whose primary focus is on models themselves. Each of these groups is heavily populated.
28. Principle 2 governing the work of the IPCC reads: 'The role of the IPCC is to assess on a comprehensive, objective, open and transparent basis the scientific, technical and socio-economic information relevant to understanding the scientific basis of risk of human-induced climate change, its potential impacts and options for adaptation and mitigation. IPCC reports should be neutral with respect to policy, although they may need to deal objectively with scientific, technical and socio-economic factors relevant to the application of particular policies.' See http://www.ipcc.ch/pdf/ipcc-principles/ipcc-principles.pdf. [accessed 16 October 2013].
29. Herman Rubin quoted in Good, *Good Thinking*.
30. D. S. Sivia, *Data Analysis: A Bayesian Tutorial* (Oxford: Clarendon Press, 2000).
31. Good, *Good Thinking*.
32. E. Mach, *Error and Uncertainty* (1856).
33. See UKCP09 web pages and presentations; for instance http://ukclimateprojections.defra.gov.uk/22612 (in web page on UKCP09 customisable maps in detail) [accessed 16 October 2013] or http://www.ukcip.org.uk/wordpress/wp-content/UKCP09/UP_0609_presentation.pdf (in slide 23 in presentation to User Panel on 3 June 2009) [accessed 16 October 2013].
34. See http://ukclimateprojections.defra.gov.uk/22530 (in web page on UKCP09 reports and guidance).
35. For an alternative view see I. Held, 'The Gap between Simulation and Understanding in Climate Modeling', *Bulletin of the American Meteorological Society*, 86 (2005), pp. 1609–14; for the reasons why we should expect difficulty in forming decision-relevant probability forecasts for systems best simulated by nonlinear models see Smith, 'What Might We Learn from Climate Forecasts?'
36. J. M. Murphy, et al., 'A Methodology for Probabilistic Predictions of Regional Climate Change from Perturbed Physics Ensembles', *Philosophical Transactions of the Royal Society A*, 365 (2007), pp. 1993–2028.
37. D. A. Stainforth, M. R. Allen, E. R. Tredger and L. A. Smith, 'Confidence, Uncertainty and Decision-Support Relevance in Climate Predictions', *Philosophical Transactions of the Royal Society A*, 365 (2007), pp. 2145–61.
38. In the same way, LeVerrier and Adams hypothesized a planet beyond Uranus to account for its motion, leading to the discovery of Neptune. In fact, LeVerrier did just this, leading to the discovery of the planet Vulcan in 1859.
39. Although applying an empirically determined statistical adjustment might lead to a model better than (i.e. more adequate than) Newton's laws on time scales, which are

much less than those on which Newton's laws (on their own) provide adequate probability forecasts for other planets.

40. Sivia, *Data Analysis*.
41. See http://ukclimateprojections.defra.gov.uk/23173 (UKCP09 science review) [accessed 16 October 2013].
42. H. H. Lamb, *Climate History and the Modern World* (Oxford: Oxford University Press, 1982).
43. L. A. Smith and D. Stainforth, 'Putting the Weather Back into Climate: On the Definition of Climate' (forthcoming).
44. See http://www.ipcc.ch/publications_and_data/ar4/wg1/en/annexes.html [accessed 16 October 2013].
45. Smith and Stern, 'Scientific Uncertainty and Policy Making'.
46. S. Manabe and R. T. Wetherald, 'The Effects of Doubling the CO_2 Concentration on the Climate of a General Circulation Model', *Journal of the Atmospheric Sciences*, 32 (1975), pp. 3–15.
47. Jenkins, et al., *UK Climate Projections: Briefing Report*, version 2, December 2010.
48. See http://ukclimateprojections.defra.gov.uk/22549 (briefing report downloads) [accessed 16 October 2013].
49. See http://ukclimateprojections.defra.gov.uk/22769 (online climate change projections report Purpose & design of UKCP09) [accessed 16 October 2013].
50. See http://ukclimateprojections.defra.gov.uk/22783 (online climate change projections report 3.3 Interpretation) [accessed 16 October 2013].
51. See n. 50, above.
52. See n. 48, above.
53. See n. 50, above.
54. See http://ukclimateprojections.defra.gov.uk/23261 (weather generator) [accessed 16 October 2013].
55. See http://ukclimateprojections.defra.gov.uk/23080 (online climate change projections report Annex 3.4.2) [accessed 16 October 2013].
56. See http://ukclimateprojections.defra.gov.uk/22980 (online climate change projections report Annex 6.2) [accessed 16 October 2013].
57. See n. 55, above.
58. See n. 55, above.
59. See n. 55, above. T. N. Palmer, F. J. Doblas-Reyes, A. Weisheimer and M. J. Rodwell, 'Toward Seamless Prediction: Calibration of Climate Change Projections using Seasonal Forecasts', *Bulletin of the American Meteorological Society* (2008), pp. 459–70; A. A. Scaife, C. Buontempo, M. Ringer, M. Sanderson, C. K. Gordon and J. Mitchell, 'Comment on Toward Seamless Prediction: Calibration of Climate Change Projections using Seasonal Forecasts' (2008).
60. K. Beven, 'On Undermining the Science?', *Hydrological Processes*, 20 (2006). pp. 3141–6.
61. Good, *Good Thinking*.
62. See n. 13, above.
63. Smith, 'What Might We Learn from Climate Forecasts?'
64. A. Stirling, 'Keep it Complex', *Nature*, 468 (2010), pp. 1029–31.
65. Arguably, every user of the probabilities for 2080 is 'exposed to' the impacts of the global model generating the 'wrong weather' worldwide, and accumulating impacts of the failure of local downscaling to simulate 'blocking' realistically, for instance, over the intervening six decades.

66. One might argue that the contents of a minority opinion should be part and parcel of the study itself. While we would agree with this goal, the facts on the ground indicate that, for whatever reason, clear information of the potential irrelevance of the finding of the study is often not highlighted in practice. A major aim in suggesting a minority opinion is to ultimately raise the profile of this information in the primary study.
67. KNMI (2012), *Advisory Board Report: Towards the KNMI '13 Scenarios – Climate Change in the Netherlands*, KNMI Publication 230 (De Bilt: Royal Netherlands Meteorological Institute), available at http://www.knmi.nl/climatescenarios/documents/AdvisoryBoard_report_towards_KNMI13.pdf [accessed 16 October 2013].
68. W. Hazeleger, B. J. J. M. van den Hurk, E. Min, G. J. van Oldenborgh, X. Wang, A. C. Petersen, D. A. Stainforth, E. Vasileiadou and L. A. Smith, 'Tales of Future Weather' (forthcoming).

7 Ben-Haim, 'Order and Indeterminism: An Info-Gap Perspective

1. See also Y. Ben-Haim, 'Doing Our Best: Optimization and the Management of Risk', *Risk Analysis*, 32:8 (2012), pp. 1326–32; and Y. Ben-Haim, 'Why Risk Analysis is Difficult, and Some Thoughts on How to Proceed', *Risk Analysis*, 32:10 (2012), pp. 1638–46; B. Schwartz, Y. Ben-Haim and C. Dacso, 'What Makes a Good Decision? Robust Satisficing as a Normative Standard of Rational Behaviour', *Journal for the Theory of Social Behaviour*, 41:2 (2011), pp. 209–27, for further non-technical discussion.
2. W. Appeltans, et al., 'The Magnitude of Global Marine Species Diversity', *Current Biology*, 22 (2012), pp. 2189–202; A. D. Chapman, *Numbers of Living Species in Australia and the World,* 2nd edn (Canberra: Australian Govt., Dept. of Environment, Heritage, Environment and the Arts, 2009); and M. J. Costello, R. M. May and N. E. Stork, 'Can We Name Earth's Species Before They Go Extinct?', *Science*, 339 (2013), pp. 413–16.
3. L. Wittgenstein, *Tractatus Logico-Philosophicus*, trans. D. R. Pears and B. F. McGuinness (Routledge, 1961), section 7.
4. S. J. Gould, *Wonderful Life: The Burgess Shale and the Nature of History* (New York: W. W. Norton, 1989), p. 212.
5. Ibid., p. 236.
6. C. S. Gray, *War, Peace and Victory: Strategy and Statecraft for the Next Century* (New York: A Touchstone Book, 1990), p. 315.
7. A. Lindgren, *Pippi Longstocking* (Oxford: Oxford University Press, 2007).
8. Y. Ben-Haim, 'Jabberwocky. Or: Grand Unified Theory of Uncertainty???' (2011), at http://decisionsand-info-gaps.blogspot.com/2011/12/jabberwocky-or-grand-unified-theory-of.html.
9. H. Adams, *The Education of Henry Adams*, ed. E. Samuels (Boston, MA: Houghton Mifflin, 1918), p. 489.
10. K. R. Popper, *The Poverty of Historicism* (New York: Harper Torchbooks, 1964 [1957]), pp. 133–4.
11. Mayo, *Error and the Growth of Experimental Knowledge*.
12. L. J. Savage, *The Foundations of Statistics*, 2nd edn (New York: Dover Publications, 1972), p. 59.
13. J. G. March, 'Bounded Rationality, Ambiguity, and the Engineering of Choice', in D. E. Bell, H. Raiffa and A. Tversky (eds), *Decision Making: Descriptive, Normative, and Prescriptive Interactions* (Cambridge: Cambridge University Press, 1988), pp. 33–57, on p. 51.

14.	F. H. Knight, *Risk, Uncertainty and Profit* (New York: Harper Torchbooks, 1965 [1921]).
15.	A. Wald, 'Statistical Decision Functions which Minimize the Maximum Risk', *Annals of Mathematics*, 46:2 (1945), pp. 265–80.
16.	R. J. Lempert, S. W. Popper and S. C. Bankes, *Shaping the Next 100 Years: New Methods for Quantitative, Long-Term Policy Analysis* (Santa Monica, CA: RAND Corp, 2003).
17.	Y. Ben-Haim, *Info-Gap Decision Theory: Decisions Under Severe Uncertainty,* 2nd edn (London: Academic Press, 2006).
18.	Y. Ben-Haim, C. D. Osteen and L. J. Moffitt, 'Policy Dilemma of Innovation: An Info-Gap Approach', *Ecological Economics*, 85 (2013), pp. 130–8.
19.	Ben-Haim, *Info-Gap Decision Theory*, ch. 11.
20.	Y. Ben-Haim and K. Jeske, 'Home-Bias in Financial Markets: Robust Satisficing with Info-Gaps', Federal Reserve Bank of Atlanta, Working Paper Series, 2003:35, pdf file on the Federal Reserve Bank website. SSRN abstract and full paper at http://ssrn.com/abstract=487585 [accessed 4 October 2013].
21.	P. A. M. Dirac, *The Principles of Quantum Mechanics*, 4th edn (Oxford University Press, 1958).
22.	Ibid., p. 6.
23.	P. M. Morse and H. Feshbach, *Methods of Theoretical Physics* (New York: McGraw-Hill, 1953), section 3.2; C. Lanczos, *The Variational Principles of Mechanics*, 2nd edn (New York: Dover Publications, 1970), p. 60; Dirac, *The Principles of Quantum Mechanics*, p. 128.
24.	R. P. Feynman, 'Space-Time Approach to Non-Relativistic Quantum Mechanics', *Reviews of Modern Physics*, 20:2 (1948), pp. 367–87.
25.	Ibid., p. 371.
26.	J. Habermas, *On the Logic of the Social Sciences*, trans. S. Nicholsen and J. A. Stark (MIT Press, 1990 [1970]).
27.	R. R. Nelson and S. G. Winter, *An Evolutionary Theory of Economic Change* (Cambridge, MA: Belknap Press, 1982), p. 370.
28.	J. M. Keynes, *The General Theory of Employment, Interest, and Money* (Amherst, NY: Prometheus Books, 1997 [1936]), pp. 198, 199, 204.
29.	G. L. S. Shackle, *Epistemics and Economics: A Critique of Economic Doctrines* (Transaction Publishers, 1992; Cambridge University Press, 1972), pp. 3–4, 156, 239, 401–2.
30.	K. R. Popper, *The Open Universe: An Argument for Indeterminism*, Postscript to *The Logic of Scientific Discovery* (London: Routledge, 1982), pp. 80–1, 109.
31.	Y. Ben-Haim, 'Peirce, Haack and Info-Gaps', in C. de Waal (ed.), *Susan Haack, A Lady of Distinctions: The Philosopher Responds to Her Critics* (Amherst, New York: Prometheus Books, 2007), pp. 150–64.
32.	Y. Carmel and Y. Ben-Haim, 'Info-Gap Robust-Satisficing Model of Foraging Behavior: Do Foragers Optimize or Satisfice?', *American Naturalist*, 166 (2005), pp. 633–41.
33.	A. Mas-Colell, M. D. Whinston and J. R. Green, *Microeconomic Theory* (Oxford: Oxford University Press, 1995).
34.	H. A. Simon, 'A Behavioral Model of Rational Choice', *Quarterly Journal of Economics*, 69:1 (1955), pp. 99–118; H. A. Simon, 'Rational Choice and the Structure of the Environment', *Psychological Review*, 63:2 (1956), pp. 129–38.
35.	Oxford English Dictionary (1989), 2nd edn.
36.	Knight, *Risk, Uncertainty and Profit*.
37.	Ben-Haim, *Info-Gap Decision Theory*.
38.	S. Chinnappen-Rimer and G. P. Hancke, 'Actor Coordination Using Info-Gap Decision Theory in Wireless Sensor and Actor Networks', *International Journal of Sensor Networks*, 10:4 (2011), pp. 177–91; D. R. Harp and V. V. Vesselinov, 'Contaminant Remediation Decision Analysis Using Information Gap Theory', *Stochastic Environmen-*

tal Research and Risk Assessment, 27:1 (2012), pp. 159–68; Y. Kanno and I. Takewaki, 'Robustness Analysis of Trusses with Separable Load and Structural Uncertainties', *International Journal of Solids and Structures*, 43:9 (2006), pp. 2646–69.

39. M. Burgman, *Risks and Decisions for Conservation and Environmental Management* (Cambridge: Cambridge University Press, 2005).
40. Y. Ben-Haim, *Info-Gap Economics: An Operational Introduction* (London: Palgrave, 2010); Ben-Haim, Osteen and Moffitt, 'Policy Dilemma of Innovation'; T. Knoke, 'Mixed Forests and Finance – Methodological Approaches', *Ecological Economics*, 65:3 (2008), pp. 590–601.
41. Y. Ben-Haim, N. M. Zetola and C. Dacso, 'Info-Gap Management of Public Health Policy for TB with HIV-Prevalence', *BMC Public Health*, 12 (2012), p. 1091.
42. L. J. Moffitt, J. K. Stranlund and B. C. Field, 'Inspections to Avert Terrorism: Robustness under Severe Uncertainty', *Journal of Homeland Security and Emergency Management*, 2:3 (2005).
43. J. W. Hall, et al., 'Robust Climate Policies under Uncertainty: A Comparison of Robust Decision Making and Info-Gap Methods', *Risk Analysis*, 32:10 (2012), pp. 1657–72.
44. Written by J. Lennon and P. McCartney (1967).
45. Carmel and Ben-Haim, 'Info-Gap Robust-Satisficing Model of Foraging Behavior'.
46. Ben-Haim, *Info-Gap Decision Theory*, section 11.4; Y. Ben-Haim, 'When is Non-Probabilistic Robustness a Good Probabilistic Bet?', working paper (2011), at http://www. technion.ac.il/yakov/prx27a.pdf [accessed 4 October 2013]; Y. Ben-Haim, 'Robust Satisficing and the Probability of Survival', *International Journal of System Science* (2012, forthcoming).
47. Ben-Haim, Osteen and Moffitt, 'Policy Dilemma of Innovation'.
48. Ibid.
49. Y. Ben-Haim, Optimizing and Satisficing in Quantum Mechanics: An Info-Gap Approach, working paper (2011), at http://www.technion.ac.il/yakov/lr05.pdf [accessed 4 October 2013].
50. S. Haack, *Manifesto of a Passionate Moderate: Unfashionable Essays* (Chicago, IL: University of Chicago Press, 1998).

8 Spanos, 'Learning from Data: The Role of Error in Statistical Modelling and Inference'

1. I would like to thank Deborah Mayo for numerous creative discussions on several issues discussed in this chapter.
2. See D. A. Redman, *The Rise of Political Economy as a Science* (Cambridge, MA: MIT Press, 1997).
3. See A. Spanos, 'Statistics and Economics', in S. N. Durlauf and L. E. Blume (eds), *New Palgrave Dictionary of Economics*, 2nd edn (London: Palgrave Macmillan, 2008), pp. 1129–62.
4. M. Blaug, *Economic Theory in Retrospect*, 5th edn (Cambridge: Cambridge University Press, 1997); Redman, *The Rise of Political Economy as a Science*.
5. JSSL, 'Prospects of the Objects and Plan of Operation of the Statistical Society of London', *Journal of the Statistical Society of London* (1869), p. 2.
6. J. S. Mill, *Essays on Some Unsettled Questions of Political Economy* (New York: A. M. Kelley, 1974 [1844]).
7. D. Ricardo, *Principles of Political Economy and Taxation, The Collected Works of David Ricardo*, vol. 1, eds P. Sraffa and M. Dobb (Cambridge: Cambridge University Press,

1817). See D. A. Redman, *Economics and the Philosophy of Science* (Oxford: Oxford University Press, 1991); Redman, *The Rise of Political Economy as a Science*.

8. Mill, *Essays on some Unsettled Questions of Political Economy*.

9. See D. M. Hausman, *The Inexact and Separate Science of Economics* (Cambridge: Cambridge University Press, 1992) for a modern interpretation of Mill's thesis.

10. J. M. Keynes, 'Professor Tinbergen's Method', *Economic Journal*, 49 (1939), pp. 558–68.

11. J. Tinbergen, *Statistical Testing of Business Cycle Theories*, 2 vols (Geneva: League of Nations, 1939).

12. M. Boumans and J. B. Davis, *Economic Methodology: Understanding Economics as a Science* (London: Palgrave Macmillan, 2010); D. F. Hendry and M. S. Morgan (eds), *The Foundations of Econometric Analysis* (Cambridge: Cambridge University Press, 1995); M. S. Morgan, *The History of Econometric Ideas* (Cambridge: Cambridge University Press, 1990).

13. L. Robbins, *An Essay on the Nature and Significance of Economic Science*, 2nd edn (London: McMillan, 1935).

14. See R. Frisch, *Statistical Confluence Analysis by Means of Complete Regression Schemes* (Oslo: Universitetets Okonomiske Institutt, 1934), p. 6.

15. Ricardo, *Principles of Political Economy and Taxation*.

16. Mayo, *Error and the Growth of Experimental Knowledge*.

17. See S. Stigler, 'Fisher in 1921', *Statistical Science*, 20 (2005), pp. 32–49.

18. R. A. Fisher, 'On the Mathematical Foundations of Theoretical Statistics', *Philosophical Transactions of the Royal Society A*, 222 (1922), pp. 309–68.

19. Ibid.; R. A. Fisher, 'Theory of Statistical Estimation', *Proceedings of the Cambridge Philosophical Society*, 22 (1925), pp. 700–25; R. A. Fisher, 'Two New Properties of Maximum Likelihood', *Proceedings of the Royal Statistical Society A*, 144 (1934), pp. 285–307.

20. J. Neyman and E. S. Pearson, 'On the Problem of the Most Efficient Tests of Statistical Hypotheses', *Philosophical Transactions of the Royal Society A*, 231 (1933), pp. 289–337.

21. J. Neyman, 'Outline of a Theory of Statistical Estimation Based on the Classical Theory of Probability', *Philosophical Transactions of the Royal Statistical Society of London A*, 236 (1937), pp. 333–80.

22. A. L. Bowley, *Elements of Statistics*, 6th edn (London: Staples Press, 1937).

23. F. C. Mills, *Statistical Methods* (New York: Henry Holt and Co., 1924).

24. R. G. D. Allen, *Statistics for Economists* (London: Hutchinson's University Library, 1949).

25. See A. N. Kolmogorov, *Foundations of the Theory of Probability*, 2nd edn (Chelsea Publishing Co., NY, 1933).

26. See J. L. Doob, *Stochastic Processes* (New York: Wiley, 1953).

27. P. Hall and C. C. Heyde, *Martingale Limit Theory and its Applications* (London: Academic Press, 1980).

28. See A. Spanos, 'A Frequentist Interpretation of Probability for Model-Based Inductive Inference', *Synthese* (2013, forthcoming).

29. See D. R. Cox and D. V. Hinkley, *Theoretical Statistics* (London: Chapman and Hall, 1974).

30. See E. L. Lehmann, *Testing Statistical Hypotheses*, 2nd edn (New York: Wiley, 1986).

31. See A. Spanos, 'Econometrics in Retrospect and Prospect', in T. C. Mills and K. Patterson (eds), *New Palgrave Handbook of Econometrics*, 00 vols (London: MacMillan, 2006), vol. 1, pp. 3–58.

32. See A. Spanos and A. McGuirk, 'The Model Specification Problem from a Probabilistic Reduction Perspective', *Journal of the American Agricultural Association*, 83 (2001), pp. 1168–76.

33. For more details see A. Spanos, 'Statistical Misspecification and the Reliability of Inference: The Simple t-test in the Presence of Markov Dependence', *Korean Economic Review*, 25 (2009), pp. 165–213.
34. Ricardo, *Principles of Political Economy and Taxation*.
35. See A. Spanos, 'Theory Testing in Economics and the Error Statistical Perspective', in Mayo and Spanos (eds), *Error and Inference*, pp. 202–46.
36. See A. Spanos, 'Curve-Fitting, the Reliability of Inductive Inference and the Error-Statistical Approach', *Philosophy of Science*, 74 (2007), pp. 1046–66.
37. See W. H. Greene, *Econometric Analysis*, 7th edn (London: Prentice Hall International, 2011).
38. See ibid., amongst others.
39. T. L. Lai and H. Xing, *Statistical Models and Methods for Financial Markets* (New York: Springer, 2008).
40. See ibid., pp. 72–81.
41. Ibid.
42. See Spanos, 'Econometrics in Retrospect and Prospect'.
43. Mayo and Spanos, 'Error Statistics', in P. S. Bandyopadhyay and M. R. Forster (eds), *Philosophy of Statistics, Handbook of the Philosophy of Science* (Oxford: Elsevier, 2011), pp. 151–196.
44. D. G. Mayo and A. Spanos, 'Severe Testing as a Basic Concept in a Neyman-Pearson Philosophy of Induction', *British Journal for the Philosophy of Science*, 57 (2006), pp. 323–57.
45. D. G. Mayo, 'Error Statistics and Learning from Error', *Philosophy of Science*, 64 (1997), pp. 195–212.
46. Mayo and Spanos, 'Error Statistics'.
47. Mill, *Essays on some Unsettled Questions of Political Economy*.
48. Keynes, 'Professor Tinbergen's Method'.
49. See diagram 1.2 in A. Spanos, *Statistical Foundations of Econometric Modelling* (Cambridge: Cambridge University Press, 1986).
50. Suppes, 'Models of Data'.
51. Called *primary* in Mayo, *Error and the Growth of Experimental Knowledge*.
52. Spanos, *Statistical Foundations of Econometric Modelling*. At the time, the author was unaware of Suppes, 'Models of Data'. The sequence of models in the diagram was inspired by T. Haavelmo, 'The Probability Approach in Econometrics', *Econometrica*, 12 (1944), supplement, pp. 1–115; see A. Spanos, 'On Re-Reading Haavelmo: A Retrospective View of Econometric Modeling', *Econometric Theory*, 5 (1989), pp. 405–29.
53. See Haavelmo, 'The Probability Approach in Econometrics'.
54. Spanos, 'Curve-Fitting, the Reliability of Inductive Inference and the Error-Statistical Approach'.
55. See Mayo and Spanos, 'Severe Testing as a Basic Concept in a Neyman-Pearson Philosophy of Induction'.
56. See A. Spanos, 'The Simultaneous Equations Model Revisited: Statistical Adequacy and Identification', *Journal of Econometrics*, 44 (1990), pp. 87–108.
57. In Spanos, *Statistical Foundations of Econometric Modelling*.
58. See Morgenstern, *On the Accuracy of Economic Observations*.
59. See A. Spanos, 'On Theory Testing in Econometrics: Modeling with Nonexperimental Data', *Journal of Econometrics*, 67 (1995), pp. 189–226.
60. See K. D. Hoover, *Causality in Macroeconomics* (Cambridge: Cambridge University Press, 2001); F. Guala, *The Methodology of Experimental Economics* (Cambridge: Cambridge University Press, 2005).

61. A. Spanos, 'Revisiting the Omitted Variables Argument: Substantive vs. Statistical Adequacy', *Journal of Economic Methodology*, 13 (2006), pp. 179–218.

62. See Greene, *Econometric Analysis* and P. Kennedy, A Guide to Econometrics, 6th edn (Cambridge, MA: MIT Press, 2008); J. M. Wooldridge, *Introductory Econometrics: A Modern Approach*, 3rd edn (London: South-Western and Thomson Learning, 2009), amongst others.

63. See Spanos, *Probability Theory and Statistical Inference*.

64. See Spanos, 'The Simultaneous Equations Model Revisited'.

65. See A. Spanos, 'Revisiting Data Mining: "Hunting" with or without a License', *Journal of Economic Methodology*, 7 (2000), pp. 231–64.

66. See A. Spanos, 'Where Do Statistical Models Come From? Revisiting the Problem of Specification', in J. Rojo (ed.), *Optimality: The Second Erich L. Lehmann Symposium*, Lecture Notes-Monograph Series, vol. 49 (Institute of Mathematical Statistics, 2006), pp. 98–119.

67. H. Leeb and B. M. Pötscher, 'Model Selection and Inference: Facts and Fiction', *Econometric Theory*, 21 (2005), pp. 21–59.

68. See A. Spanos, 'Akaike-Type Criteria and the Reliability of Inference: Model Selection vs. Statistical Model Specification', *Journal of Econometrics*, 158 (2010), pp. 204–20.

69. E. F. Fama and K. R. French, 'Multifactor Explanations of Asset Pricing Anomalies', *Journal of Finance*, 51 (1996), pp. 55–84.

70. H. Levy, *The Capital Asset Pricing Model in the 21st Century: Analytical, Empirical, and Behavioral Perspectives* (Cambridge: Cambridge University Press, 2011).

71. J. M. Keynes, 'Comment', *Economic Journal*, 50 (1940), pp. 155–6. The original quotation from Keynes is: 'It will be remembered that the seventy translators of the Septuagint were shut up in seventy separate rooms with the Hebrew text and brought out with them, when they emerged, seventy identical translations. Would the same miracle be vouchsafed if seventy multiple correlators were shut up with the same statistical material?'

72. Spanos, 'Revisiting the Omitted Variables Argument'.

73. See Spanos, 'Curve-Fitting, the Reliability of Inductive Inference and the Error-Statistical Approach'.

74. D. G. Mayo and A. Spanos, 'Methodology in Practice: Statistical Misspecification Testing', *Philosophy of Science*, 71 (2004), pp. 1007–25.

75. See Spanos and McGuirk, 'The Model Specification Problem from a Probabilistic Reduction Perspective'.

76. Like those in G. U. Yule, 'Why Do We Sometimes Get Nonsense Correlations Between Time Series-A Study in Sampling and the Nature of Time Series', *Journal of the Royal Statistical Society*, 89 (1926), pp. 1–64.

77. See Spanos, 'Curve-Fitting, the Reliability of Inductive Inference and the Error-Statistical Approach'.

78. Mayo and Spanos, 'Severe Testing as a Basic Concept in a Neyman-Pearson Philosophy of Induction'.

79. See Lehmann, *Testing Statistical Hypotheses*.

80. Mayo and Spanos, 'Severe Testing as a Basic Concept in a Neyman-Pearson Philosophy of Induction'.

81. I. C. W. Hardy (ed.), *Sex Ratios: Concepts and Research Methods* (Cambridge: Cambridge University Press, 2002).

INDEX